Media, Environment and the Network Society

Palgrave Studies in Media and Environmental Communication

Global media and communication processes are central to how we know about and make sense of our environment and to the ways in which environmental concerns are generated, elaborated and contested. They are also core to the way information flows are managed and manipulated in the interest of political, social, cultural and economic power. While mediation and communication have been central to policy-making and to public and political concern with the environment since its emergence as an issue, it is particularly the most recent decades that have seen a maturing and embedding of what has broadly become known as environmental communication.

This series builds on these developments by examining the key roles of media and communication processes in relation to global as well as national/local environmental issues, crises and disasters. Characteristic of the cross-disciplinary nature of environmental communication, the series showcases a broad range of theories, methods and perspectives for the study of media and communication processes regarding the environment. Common to these is the endeavour to describe, analyse, understand and explain the centrality of media and communication processes to public and political action on the environment.

Titles include:

Alison G. Anderson
MEDIA, ENVIRONMENT AND THE NETWORK SOCIETY

Stephen Depoe and Jennifer Peeples (*editors*)
VOICE AND ENVIRONMENTAL COMMUNICATION

Palgrave Studies in Media and Environmental Communication
Series Standing Order ISBN 978–1–137–38433–1 (hardback)
978–1–137–38434–8 (paperback)
(*outside North America only*)

You can receive future titles in this series as they are published by placing a standing order. Please contact your bookseller or, in case of difficulty, write to us at the address below with your name and address, the title of the series and the ISBN quoted above.

Customer Services Department, Macmillan Distribution Ltd, Houndmills, Basingstoke, Hampshire RG21 6XS, England

---

Media, Environment and the Network Society

Alison G. Anderson
University of Plymouth, UK

First published 2014 by
PALGRAVE MACMILLAN

Palgrave Macmillan in the UK is an imprint of Macmillan Publishers Limited, registered in England, company number 785998, of Houndmills, Basingstoke, Hampshire RG21 6XS.

Palgrave Macmillan in the US is a division of St Martin's Press LLC, 175 Fifth Avenue, New York, NY 10010.

Palgrave Macmillan is the global academic imprint of the above companies and has companies and representatives throughout the world.

Palgrave® and Macmillan® are registered trademarks in the United States, the United Kingdom, Europe and other countries.

ISBN 978–0–230–21761–4

This book is printed on paper suitable for recycling and made from fully managed and sustained forest sources. Logging, pulping and manufacturing processes are expected to conform to the environmental regulations of the country of origin.

A catalogue record for this book is available from the British Library.

A catalog record for this book is available from the Library of Congress.

To my parents, Brian (who sadly passed away during the writing of this book) and Jean

Contents

Figures and Tables

Figures

Tables

Acknowledgements

This book would not have been possible without the input of a number of people. I am indebted to the journalists, scientists, environmental activists and policymakers who freely gave up their time to be interviewed. I would also like to thank Alan Petersen for his steadfast support and enthusiasm. Chapter 6 draws upon joint work on nanotechnologies, the media and stakeholder perceptions funded by the Economic and Social Research Council RES-000-22-0596 (with Alan Petersen, Stuart Allan and Clare Wilkinson) and the British Academy SG-44284 (with Alan Petersen and Rachel Torr). Colleagues in the School of Government, particularly research student Tim Ingram, and third year undergraduate students in 'Media, State and Society' have helped me to formulate ideas for the book. Mel Joyner was a fantastic source of support. I also greatly benefited from the invitation to present my work at the Wageningen University Communication Science seminar series in 2011 and for the lively discussion this generated. I would also like to formally acknowledge the Faculty of Social Science and Business (now the Faculty of Business) and the University of Plymouth for their generous support while working on this project. At Palgrave, Chris Penfold (initially Catherine Mitchell), and colleagues have been brilliant. I would also like to gratefully acknowledge the interest and support of the Series Editors, Anders Hansen and Steve Depoe. Finally especial thanks to my partner, Peter, for putting up with the many hours I spent away at my desk and for providing invaluable advice, criticism and encouragement, and to my son, James, whose enthusiasm for digital media ensured this book kept pace with this rapidly changing field.

1
Introduction

As I write this book Super Typhoon Haiyan, thought to be the strongest storm ever recorded, has devastated large areas of the Philippines and is thought to have killed over 10,000 people. Yet the science linking tropical storms and climate change is unclear. The most recent Intergovernmental Panel on Climate Change (IPCC) report, released in September 2013, found that:

> Globally, there is *low confidence* in attribution of changes in tropical cyclone activity to human influence. This is due to insufficient observational evidence, lack of physical understanding of the links between anthropogenic drivers of climate and tropical cyclone activity, and the low level of agreement between studies as to the relative importance of internal variability, and anthropogenic and natural forcings.
>
> (IPCC, 2013, original emphasis)

With the exception of the North Atlantic region (where it is thought to have contributed, at least in part, to increased tropical cyclone activity there since the 1970s) there is simply not enough scientific evidence to support a link. But that does not necessarily rule out a connection; it is extremely difficult to attribute specific extreme weather events to climate change. The IPCC report concludes that the weight of evidence for anthropogenic (human-caused) climate change is mounting:

> The evidence is stronger that observed changes in the climate system can now be attributed to human activities on global and regional scales in many components...Taken together, the combined evidence increases the overall level of confidence in the attribution of

observed climate change, and reduces the uncertainties associated with assessment based on a single climate variable. From this combined evidence it is *virtually certain* that human influence has warmed the global climate system.

(IPCC, 2013, original emphasis)

The report also notes that the heavy rainfalls associated with tropical cyclones and average tropical cyclone maximum wind speed are both likely to increase in the future with the continued warming of the oceans.

While the seriousness of the threat, the rate of change and the links to the frequency and intensity of extreme weather events are the subject of much debate, there is a clear scientific consensus that the mean global temperature has grown significantly since pre-industrial times and human activity is the principal cause (see Brulle et al., 2012). Yet climate change has become a highly politicised issue and the media have played a central role in shaping public perceptions and policy agendas. This book argues that analysis of media representations of environmental issues, including climate change, must be placed in the wider context of the increasing concentration and globalisation of news media ownership, and an increasingly 'promotional culture', highlighted by the rapid rise of the public relations industry in recent years and claims-makers who employ increasingly sophisticated media strategies.

One of the greatest challenges of our time concerns our relationship with nature and the environment. The choices we make now will affect generations to come. No longer conceived of as purely scientific issues, problems such as climate change, energy depletion, species extinction, deforestation, population growth, water scarcity and air pollution have permeated into public discourse and popular media over the last couple of decades. Media coverage has tended to focus more on the problems than the solutions. Many environmental issues are complex, uncertain and involve long time spans. These sit uneasily in news media schedules that favour certainty, immediacy and simplicity. Emerging technologies such as nanotechnologies and synthetic biology herald a new world that proponents argue potentially offer a means of solving some environmental problems, but at the same time some environmentalists argue they may unleash harmful effects on nature. Given so many uncertainties some social actors advocate a 'wait and see' approach while other claims-makers recommend the precautionary 'better safe than sorry' principle.

As important as studying the role of the media is examining the struggle among news sources to define issues and control agendas. Various social actors including scientists, industry, policymakers and non-governmental organisations (NGOs) battle to influence public perceptions. How the problems and solutions are framed has the potential to critically shape political and public responses. Environmental issues are often deeply contested. Media representations do not mirror objective indicators of environmental damage but offer a highly socially constructed version of reality. Also, in tracking their prominence, this book suggests we need a longitudinal, multi-issue approach given that environmental issues may compete against one another for news attention as well as competing with other non-environmental issues. For example, US news media attention was focused for much of 2010 on the Deepwater Horizon crisis, but a study just focusing on climate change coverage during that period might have concluded that environmental issues were very low down the agenda.

This book is about power – the power to influence news media agendas and propel issues onto the agenda. But there is also a less visible side to power that is about silencing issues and the ability to keep them off the radar. Among the key questions it seeks to address are: What factors trigger particular environmental stories to make their way into the headlines and others to be ignored? How do issue-attention cycles operate? Whose voices tend to have the greatest prominence in media coverage? How do actors seek to keep issues off the agenda? Why do some claims-makers choose not to be visible in mainstream media? And has the digital revolution fundamentally altered the balance of power? To what extent are new social media simply acting as echo chambers? How influential are global news agencies/wire services in the recycling of news agendas? And how does this vary across different issue domains and in different cultural contexts?

Since I wrote *Media, Culture and the Environment* (published in 1997) the media landscape has considerably changed and trends such as the reliance of environmental journalists on a show business type of approach to covering the environment have intensified. The field itself has considerably grown, so much so that *environmental communication* has become an area of multidisciplinary study in its own right (see Cox, 2013). There is now greater recognition of how values, ideology, discourse and the symbolic realm influence how we perceive nature. This book examines the influential theory of the 'network society' and discusses its significance for understanding the nature of contemporary environmental activism and mediation of the environment.

With the growing fragmentation of the media, digitalisation and a proliferation of outlets, it is hard to identify distinct boundaries between online and offline media. Web 2.0, the so-called second generation of web-based technologies, has enhanced the interactive and participatory features of the internet through blogs, wikis, video sharing, social networking and podcasting. More recently Web 3.0 (sometimes referred to as the semantic web) enables the integration of data from diverse sources through, for example, the smart tagging of documents, images, events and locations, making internet searches more precise and tailored to individuals. The widespread use of inexpensive but reasonably high quality mobile phone and digital cameras has led 'publics' to take a greater role in newsgathering through 'citizen journalism', playing a more active role in processes of news production. Increasing numbers of people are uploading their video clips of news events to media outlets (Allan, 2006; Anderson, 2006). Round the clock 24/7 reporting increases the immediacy of breaking news, accelerating the speed at which politics is conducted. RSS (Really Simple Syndication) enables users to filter news headlines according to their own personal interests. By the end of 2014 it is estimated that there will be almost 3 billion internet users worldwide (Voice of America, 2014). And advances in technology now mean that many mobile phone users can access news updates directly via their phone. Blogging has taken off to such an extent that it is estimated that there are over 175,000 new blogs posted each day (Technorati, 2008). The total number of smartphones is estimated to reach over three billion by 2017, with particularly rapid increases predicted to occur in Africa and the Middle East.

The internet itself is far from new; the technology dates back to 1969, when the first message was sent over ARPANET, and in 1971 the first network email system was created. British software engineer, Tim Berners-Lee, invented the World Wide Web (WWW) in 1989 and by the summer of 1991 his browser became available on the internet and the first web page was published. From the very beginning the environmental movement has been at the forefront of developments in new media. One of the first internet-based media projects was GreenNet in the UK (see http://www.gn.apc.org/). This, together with PeaceNet in the US, was established in the mid 1980s (Atton, 2007). In the early days pressure groups such as Greenpeace were reliant upon the use of telex and fax. Greenpeace's first email connection was made in the United States in 1983, and by the late 1980s many environmental groups were using email and newsgroups (see Greenpeace International Archives; Pickerill, 2006).

During the early 1990s activist mailing lists began to emerge and one of the first alternative media sites to utilise the benefits of the World Wide Web was the centre for environmental information, EnviroWeb. Greenpeace launched its first website in 1994 (Greenpeace International Archives). In the same year UK Friends of the Earth (FoE) established its email system and website. The website was developed by a couple of individuals in their free time, bypassing formal approvals. Later, when funding was needed, this development was stalled for a time by bureaucracy and organisational policy constraints (Pickerill, 2004). To begin with many FoE staff did not have individual access to computers and few of the regional groups actively used emails to connect with each other. The computers were slow and mainly used by FoE for membership database support, and network facilities were often unreliable (Washbourne, 2001). However, by 1996 there was one computer to each full-time member of staff based at the national headquarters and seven regional offices. The computers provided employees with a variety of facilities including email and internet access, and the group purchased a number of notebook computers to enable campaigners to keep in contact while mobile (Washbourne, 2001). In 1995 FoE launched its Chemical Release Inventory (CRI) which enabled the public to perform user-friendly postcode searches on the website for details of polluting factories in their local area, using previously incomprehensible data provided by the Environment Agency (see Pickerill, 2003).

By 1998 the website was getting more than 20,000 hits a week (FoE Annual Review 1998, cited by Pickerill, 2001: 150). The success of the CRI led to the Factory Watch campaign, established in 1999. In November 2000 Greenpeace International launched its own Cyberactivist Community, an action and discussion forum (see http://activism.greenpeace.org/cybercentre/ and http://forum.greenpeace.org/int/). Its developers describe it as providing a 'cyberactivist community where people representing over 170 countries and territories can share ideas and participate in environmental actions...' (Greenpeace, 2006). Greenpeace also facilitated cyberactivism in countries with very limited access to computers or telephone lines. In 1999 it established a cyber-café at the site of the Union Carbide Factory in Bhopal, India, where a gas leak in 1984 killed large numbers of people and left many with permanent injuries. This enabled residents to send thousands of electronic messages to Union Carbide and the Indian Government, insisting action should be taken over the continued leaking of toxic chemicals into the local groundwater (Greenpeace, 2006). As Pickerill notes, 'Their use of the technologies has evolved from a few basic websites to an international network of email lists, contacts and the use of online tactics such as hactivism' (2006: 267). More recently, blogs, wikis and mobile phones have been increasingly utilised in environmental actions. A Canadian study investigating environmental NGO's use of virtual and physical means of influencing climate policy found that the vast majority of the 100 groups interviewed (sample drawn from environmental NGOs attending the UN Framework on Climate Change, Montreal, December 2005) used websites, email and listservs in their organisation, and over a third used blogs and online petitions (Sieber et al., 2006).

The ability to connect online and through mobile devices radically changes the physical space of social action and potentially allows it to become increasingly interactive. We are able to both receive and create media content. New connections can be formed and link to action elsewhere in a series of infinite expansions. As Nick Couldry (2012) points out, the new digital revolution is comparable to the print revolution even though the speed with which it has occurred has been infinitely faster. However, as we shall see, there are still considerable uncertainties about its effects. In the rich West we are 'supersaturated with media' that are converging at greater and greater speeds (Couldry, 2012). For example, radio and newspapers are increasingly moving online, cable and satellite channels continue to multiply, television programmes invite us to comment via Twitter and social networking sites enable us to link to other media. However, there are multiple digital divides (influenced by age, ethnicity, gender and socio-economic factors) and huge variations across the globe in terms of who has access and the ability to make their voice heard. For example, in large areas of Africa, South Asia and Latin America smart phones are rare and signals are weak. Rural, uneducated women in developing countries tend to be excluded from such forms of networking. In countries such as Niger less than one per cent of people have broadband subscriptions or mobile broadband (see Broadband Commission, 2012).

Inequalities of visibility (of issues and actors) become fundamental to understanding the operation of power in the network society. The approach this book takes is to avoid the 'media centrism' that has so often characterised research in this area. It is only through widening the lens to examine processes of news production and consumption that we develop a more complete understanding of how environmental issues are mediated.

Outline of chapters

This book is divided into the following chapters.

Chapter 2, 'Environmental Risks, Protest and the Network Society', places the media politics of the environment within the wider context of debates concerning the role of contemporary media in communicating risk. In recent years a number of scholars have highlighted the lack of sustained, rigorous analysis of the role of the media in reporting risk. This chapter provides a critical survey of the literature in the field and teases out key conceptual and methodological issues. It offers an in-depth examination of Beck's influential theory of the 'risk society' and Castells' theory of the 'network society'. It argues that analysis of media reporting of environmental risks must be placed within the broader context of the growing concentration and globalisation of news media ownership, the convergence between old and new media, and the rapid rise of the PR industry.

Chapter 3, 'News Agendas, Framing Contests and Power', begins by examining the concepts of 'framing' and 'agenda-building' in order to explore the selective nature of environmental coverage. Environmental issues may be framed in a number of different ways, depending upon the influence placed on such factors as economic development or biodiversity. Different stakeholders have varying degrees of power in commanding news media attention and being treated as credible news sources. The chapter examines whose views tend to gain visibility and considers the changing strategies of NGOs in relation to traditional news media and online news.

Chapter 4, 'The Climate Change Controversy', surveys research conducted in a range of international contexts on the role of the news media in covering climate change. This chapter explores how climate change has been framed over time and which voices have been treated as legitimate and authoritative sources. It examines the reasons why these developments have occurred and considers the impact of political agendas, public pressure and the activity of NGOs. Important questions are raised concerning objectivity and trust in the communication of controversial science.

Chapter 5, 'Oil Spills and Crisis Communication', examines the news media representation of oil spills through focusing in particular on the BP Deepwater Horizon disaster in 2010. Oils spills are often considered to be particularly newsworthy by the news media. Images of visually appealing animals such as seals make for emotive coverage that is seen as engaging the audience. However, not all major oil spills receive great publicity and media coverage is often not in proportion to the total amount of damage incurred. Today the media politics of oil spills has to be considered in the context of a rapidly changing global communications environment where many news sources have developed increasingly sophisticated strategies for targeting media and shaping news agendas. The internet, particularly for activists, is increasingly providing a key source of alternative first-hand images and narratives that challenge official accounts.

Chapter 6, 'Emerging Technologies', discusses emerging environmental issues linked to nanotechnology and biotechnology, and considers the lessons that have been learnt from the reporting of previous controversies, such as genetically modified organisms. It explains the nature of debates concerning these issues and discusses the potential role of the media in representing potential benefits and risks. This discussion draws upon the sociology of expectations, stigma, attention cycles and framing theory. While there are serious concerns over the environmental effects of such technologies, they are seen to form an integral part of the solution to many environmental problems.

Finally **Chapter 7**, 'Future Directions', brings together the main arguments of the book and offers some wider reflections on the field before highlighting future avenues for research. I suggest that we need to develop a more nuanced account of the role of the media in communicating environmental issues that takes into account the complex interplay of actors and issues competing across a range of arenas including: the media, parliament, regulatory institutions, interest and pressure groups, scientific communities, and industry. Such an approach needs to be able to account for shifting dynamics and a myriad of socio-cultural factors that may shape the problematisation of an issue in different ways. The influence of actors and arenas shifts over time, and regulatory institutions play a key role in early screening and determining which interests get represented through established policy-making channels and which may be allowed to get back-door access. This directs attention to the less visible aspects of news production processes and the hidden faces of power, given that control over the media is as much to keep certain issues marginalised or hidden as it is to publicise them. How far is the internet providing a means to bypass traditional gatekeepers and who is listening anyway? What lessons can we learn from past controversies that can inform a better understanding of the media politics of the environment?

2
Environmental Risks, Protest and the Network Society

> ... thanks to the effective merging of the on and offline, massive gatherings of people attempting to change the order of the world around them is now the new normal.
>
> Jurgenson (2012)

> It's clearer than ever: climate change is real, humans are the cause, and we have to act. Sometimes the riskiest decision you can make is to do nothing.
>
> Richard Branson (2013)

Heralded as Greenpeace's largest campaign yet, 'Save the Arctic', which was launched in June 2012, employed a combination of strategies designed to skilfully attract online and offline attention to the global oil giant Shell's exploitation of Antarctica. Visual stunts are by no means new, but the combination of theatricality, drama, high profile celebrities, parody and satire communicated to global audiences via the web together with mainstream media, illustrates a new spectacularisation of the environment characteristic of power politics in late modern society. The contest for power is more visibly played out on the media stage than at any time before.

Tactic one: Use visual symbols with high emotional resonance

Activists dressed up as homeless 'polar bears' wander lost in major cities around the globe. They provide a photogenic opportunity for television news film crews and press photographers and are well attuned to news values. Team this up with videos posted on YouTube involving well known actors and music stars. The video *Vicious Circle* is narrated by actor John Hurt with music from Icelandic composer Johann Johannsson. A further video depicts a polar bear wandering around London lost in an unfamiliar habitat with voiceovers from Jude Law and accompanying music from the band Radiohead.

Tactic two: Create a highly visual event

A fake gala party celebrating oil-drilling expansion in the Arctic, supposedly organised by Shell, takes place in Seattle. A large number of actors are brought in by Greenpeace and the Yes Men to join local energy leaders invited to the event. The centrepiece of the party is an ice sculpture resting on a bath of cola beside a scale model of a drill rig. As the guest of honour turns on the drinks dispenser that is shaped like an oilrig there is an uncontrolled blowout and black 'oil' spurts out over the crowd. The video of the fake press conference goes viral.

Tactic three: Exploit digital media to their full potential

A dedicated website http://www.savethearctic.org/ is created. This is joined by a Facebook page and Twitter page #SavetheArctic. Mobilise followers through these means.

Tactic four: Hit the web with satirical comment exposing your adversary's less sophisticated grasp of social media

Greenpeace creates an almost identical version of Shell's own website called Arctic Ready. Visitors to the website can play a game, 'Angry Bergs', a parody of the popular game 'Angry Birds'. A fake Twitter feed from Shell Social Media team, @ShellisPrepared, is created which appears to be trying to contain the negative advertisements from a social media tool on their site. The parody video quickly hits the top spot on Reddit and the number two spot on YouTube, with half a million views in less than 24 hours. The Yes Lab sent out a hoax press release supposedly from Shell in order to generate additional media coverage, threatening anyone who reposts the video and attacking the new ArcticReady.com website (Greenpeace, 2012).

Tactic five: Turn flak on its head

A series of fake legal threats are sent to journalists and bloggers warning that: 'Shell is monitoring the spread of potentially defamatory material on the Internet and reporters are advised to avoid publishing such material' (Robbins, 2012). Flak refers to negative comments/feedback to the media (including legal threats) that can dissuade certain material from being publicised. Some sections of the media are duped by the fake press release and report the Shell launch party as fact. Approximately 24 hours after the launch of the campaign Greenpeace and the Yes Men post a video along with blogs admitting their involvement. Greenpeace are accused by some of media manipulation (see Robbins, 2012). However, when many of those news outlets and blogs report that it turned out to be a hoax they gain many additional page views, so it could be seen as having benefitted both the media and environmentalists.

Tactic six: Create popular legitimacy by gaining the backing of high profile celebrities

A string of celebrities appealing to different generations including Hollywood actors and business leaders are recruited to endorse the campaign. Stars such as Robert Redford, Paul McCartney and Penelope Cruz, and bands such as Radiohead and One Direction are amongst the first 100 to be written on an Arctic Scroll planted on the seabed at the North Pole, 4 km beneath the ice and marked by a 'flag of the future'. This powerfully symbolises their support and the celebrity hook leads to coverage in popular mainstream media outlets (see Anderson, 2013).

None of these methods are new. However, online media have increased the speed and intensity with which messages can be circulated and enabled environmentalists to co-ordinate actions around the world with much more ease. There is now greater appreciation that we face a global environmental crisis that requires global action. Moreover, as Douglas Kellner observes: 'Every form of culture and more and more spheres of social life are permeated by the logic of the spectacle' (2003: vii).

Various over-arching grand theories have been developed in an attempt to capture the broad contours that shape these shifts in contemporary society. Two of the most influential theories are 'risk society' and the 'network society'. For German sociologist Ulrich Beck we are living in a 'risk society' that faces global catastrophes that are qualitatively different from earlier threats and for Spanish sociologist Manuel Castells the 'information society' has been replaced by what he calls the 'network society'. Although their focus and analyses differ they both agree that the nature of the kind of society we live in has fundamentally changed and that the media are playing an increasingly central role.

This chapter seeks to place the media politics of the environment within the wider context of debates concerning the role of contemporary media in communicating risk. In getting a fix on the role of the media in reporting environmental risks, it provides a critical survey of this contested field of research and outlines the key paradigms and perspectives. It offers an in-depth examination of Beck's influential theory of the 'risk society' and Castells' theory of the 'network society'. It argues that examination of media coverage of environmental risks needs to be placed in the wider context of the increasing globalisation and concentration of news media ownership, the rapid rise of the PR industry, and the convergence between 'old' and 'new' media. The chapter concludes by considering the implications for developing a more sophisticated account of the role of the media in communicating contemporary risks within the network society.

Risk society

In a climate of heightened public sensitivity about risk, following a series of heavily publicised disasters, Ulrich Beck's ground-breaking book *Risk Society: Towards a New Modernity* (1992) struck a chord. The text became highly influential, being translated into over 80 languages and sparking off considerable academic and political debate (see Mythen, 2007). Beck's powerful critique of industrial society suggests that modernisation and technological innovation have produced a series of damaging and increasingly unmanageable side effects. For Beck 'risk society' is: 'A phase of development of modern society in which the social, political, ecological and individual risks created by the momentum of innovation increasingly elude the control and protective institutions of industrial society' (Beck, 1994: 27). He argues that, compared with 'natural hazards' in pre-industrial society (such as drought, famines and plagues), contemporary humanly produced 'manufactured risks' (such as chemical accidents, nuclear power and environmental pollution) are more unpredictable and potentially catastrophic. They only acquire 'visibility' when defined as social problems through science, the legal apparatus and the media (see Cottle, 1998). 'What eludes sensory perception', writes Beck, 'becomes socially available to "experience" in media pictures and reports. Pictures of tree-skeletons, worm-infested fish, dead seals (whose living images have been engraved on human hearts) condense and concretize what is otherwise ungraspable in everyday life' (Beck, 1995: 100). Finally, they impact far beyond a specific geographical locale, or place in time, and possess the capacity to position humankind on the brink of annihilation. Thus he writes:

The modern world increases the worlds of difference between the language of calculable risks in which we think and act and the world of non-calculable uncertainty that we create with the same speed of its technological developments. With the past decisions on nuclear energy and our contemporary decisions on the use of genetic technology, human genetics, nanotechnology, computer sciences and so forth, we set off unpredictable, uncontrollable and incommunicable consequences that endanger life on earth.

(2002: 3)

Risk society has produced an increasing array of global environmental dangers associated with nuclear power, chemical accidents and environmental pollution that have cumulative effects over time. This is not to suggest that risk society is any more hazardous than in the past, but we have a greater orientation towards the future and a growing consciousness or awareness of risk and our own role in causing these problems. But as technological development grows apace with myriad, unpredictable consequences, institutions have difficulty in calculating and insuring against risk – raising serious issues of regulation and accountability:

The entry into risk society occurs at the moment when the hazards which are now decided and consequently produced by society *undermine and/or cancel the established safety systems of the provident state's existing risk calculations*. In contrast to early industrial risks, nuclear, chemical, ecological and genetic engineering risks (a) can be limited in terms of neither time nor place, (b) are not accountable according to the established rules of causality, blame and liability, and (c) cannot be compensated or insured against. Or, to express it by reference to a single example: the injured of Chernobyl are today, years after the catastrophe, not even all *born* yet.

(1996: 31, original emphasis)

Paradoxically, Beck contends, the more we depend upon 'experts' to manage and control risks, the more our trust in them weakens; a relationship characterised by ambivalence: 'the risk consciousness of the afflicted, which is frequently expressed in the environmental movement, and in criticism of industry, experts and culture, is usually both critical and credulous of science' (Beck, 1992: 72). These same experts are seen as relying upon the media to transmit information about risk. Beck accords the media a central role in the public understanding of risk, though as we shall see, his theory is relatively undeveloped.

For Beck, the media can act as a vehicle through which experts relay institutional information to publics and, paradoxically, as a channel for reflexivity and public critique. Particularly pertinent here is his concept of 'relations of definition', which refers to the institutions involved in assessing and managing risk including government, scientists, the legal profession and the civil service, as well as the 'cultural matrix in which risk politics is conducted' (Beck 1997b cited in Cottle, 1998: 7). These inform and shape the judgements made by 'experts', 'counter-experts' and members of the 'lay public' about the harmfulness of products, the dangers posed by particular threats and who can be held to blame. The analysis of risk, he argues, must account for the media's structurating significance in the formation of public attitudes about risk. Risks, as Beck (1992: 23) suggests, can 'be changed, magnified, dramatized or minimized within knowledge, and to that extent they are particularly *open to social definition and construction*' (original emphasis). The media, it is contended, play a central role in playing out this social contest over risk and simultaneously in stimulating social criticism. This raises key questions concerning which voices in the media come under this spotlight, under what conditions, and where it is (and is not) directed and why (Allan, 2000; Anderson, 2006). How news actors define the risks and who is held accountable becomes of crucial importance.

Beck continues to see the media as a powerful spotlight for revealing the 'side effects' of industrial capitalism in his book *World Risk Society* (1998). And in a more recent publication, *Cosmopolitan Vision* (2006), Beck argues that since the 1990s we have increasingly come to view ourselves as part of a globally connected society confronted by global threats and crises aided by transnational communication flows. Media and communications have become increasingly constitutive of the global public sphere in societies that are rapidly globalising and mediatized:

> Thus, insofar as a global everyday existence becomes an integral part of media worlds, a kind of globalization of emotions and empathy occurs. People experience themselves as parts of fragmented, endangered civilization and civil society characterised by the simultaneity of events and the knowledge of this simultaneity all over the world.
>
> (2006: 42)

Beck's account of the media has been criticised as unevenly developed, overly simplistic and in places contradictory (Cottle, 1998: 25; Mythen, 2004: 76). In writings such as *Risk Society* (1992) and *Ecological Politics in an Age of Risk* (1995) he shows little awareness of mainstream media theory (see Cottle, 1998; Murdock et al., 2003; Mythen, 2004). While Beck usefully draws our attention to several broad features of risk in contemporary society, he presents a passionately argued and politically charged abstract, macro theory of risk which is lacking in empirical detail. Clearly its overall purpose needs to be borne in mind when assessing its utility. As Mythen observes: 'The risk society thesis is assembled in the spirit of exploration and adventure: it is not driven by empirical validity, but by invigorating sociology and providing thought-provoking reflections on the modern condition' (2007: 803). Nevertheless, in order to build upon his insights we need to address the empirical gaps and theoretical weaknesses. We shall concentrate here upon the deficiencies that relate to the analysis of the media in communicating environmental risk. This is no simple task since his observations about the role of the media have to be pieced together as they tend to be scattered across a number of different publications and are frequently ambiguous (Cottle, 1998).

First Beck tends to treat the media as monolithic – that is to say, he lumps media together rather than distinguishing between different outlets and genres (Anderson, 1997: 188). Research in media and cultural studies has shown that the news media are differentiated occupying their own distinctive market niches, and governed by a range of economic and political constraints (Anderson, 1997). Representation of environmental risks varies according to different media formats that are governed by their own particular restrictions and practices. For example, risk reporting in the red-top UK newspapers tends to concentrate more on 'human interest' stories involving ordinary people's experiences than their more high-brow equivalents (Murdock et al., 2003; Petts et al., 2001) and television news stories tend to be attracted towards visually appealing items (Anderson, 1997). A study conducted by Hansen found significant differences between UK and Danish environmental coverage on two major television news programmes; the degree of relative attention to environmental issues was closely connected with the economic base of each respective country (Hansen, 1990). Further layers of complexity may also be observed when considering the variations between local and national/international coverage of environmental affairs, and the differentiated nature of environmental coverage across a range of different television programme formats from the 'popular' to the 'serious' (see Anderson, 1997; Cottle, 1993; Nohrstedt, 1991).

Second, research demonstrates the need to distinguish between how media cover different risks, given that some risks resonate more strongly with individuals in particular cultural contexts than others (Hansen, 1991; Kitzinger, 1999). For example, nuclear issues connect with powerful symbolic imagery and deep-seated fears associated with atomic bombs and radiation (Hansen, 1991). This goes some way to explain why nuclear issues tend to get more coverage than some other environmental issues which are perceived to be less threatening (Anderson, 1997).

Third, there are also important cross-cultural differences, yet Beck devotes scant consideration to distinctive national and political contexts and cultures (Cottle, 1998: 17–18). Chapman et al.'s comparison of the representation of environmental issues in the UK and India reveals significant differences – most notable of which is that in India it was very difficult to discern a separate category 'environment', singled out from development and civic issues (Chapman et al., 1997). Boykoff (2007a) also noted major differences in the reporting of climate change between US and UK elite newspapers (see Chapter 4).

Fourth, as Cottle notes, Beck's narrow focus upon the media spotlight directs attention away from examining news production processes and is: 'conspicuously silent ... on the institutional field in which "relations of definitions" compete for public recognition and legitimation' (Cottle, 1998: 18). There is therefore considerable scope for rigorous empirical analysis of source–media relations and theorisation that adequately accounts for the complexity of the role of the media in framing risk (see Chapter 3).

Finally, Beck makes strong claims about media effects without providing empirical evidence (Anderson, 1993: 51). A vast amount of research in media and cultural studies has demonstrated that, while the news media have the ability to influence public and policy agendas, there is no simple, direct relationship between public attitudes and media exposure (see Philo, 1999). The 'public' is made up of several different sub-groupings who are likely to exhibit different responses depending upon factors such as the particular risk issue in question, as well as their own individual biographical and social histories (Hornig, 1993; Hornig Priest, 2008; Tulloch & Lupton, 2001).

In comparison, Castells' work represents a more fully developed perspective on the shifting role of the media in contemporary society. It is more nuanced and less conjectural, and gives greater emphasis to the dynamic agency of actors aided by the technologies that make global networked communication possible. It is to this that we now turn.

The network society

Castells' account of the 'network society' has particular analytical value for the present discussion. Networks, he argues, are very old types of social organisation present throughout human history, but with globalisation and the weakening power of the nation state they have taken on a new boundless form, powered by micro electronics based information technologies. However, he is by no means the first to use the term 'network society' in relation to communication networks (see also van Dijk, 1991). Like Beck, Castells views the media as occupying a central role within contemporary society: 'In the network society, culture is by and large embedded in the processes of communication, in the electronic hypertext, with the media and the Internet at its core' (2004: 32–3). Elsewhere he contends: 'What does not exist in the media does not exist in the public mind, even if it could have a fragmented presence in individual minds' (2007: 241). Network society is defined as 'a social structure constructed around (but not determined by) digital networks of communication'. These 'global networks that structure all societies' are seen as lying at the heart of early twenty-first century life (2009: 53). It is in this space that political contests are fought and the ability of networks to influence these processes is crucial to their success in making their demands heard. Politicians and social movements alike have had to adapt to the new communications environment where a sound bite, celebrity-focused culture prevails. The networks are 'programmed' by powerful social actors but under certain conditions they can be 'reprogrammed' by those with alternative interests in a process of 'counter-power'.

In *Communication Power*, Castells elaborates upon some of his earlier writings and provides in general a highly insightful account of the digital revolution. In essence the network society thesis maintains that power is located within the networks that structure society, not in institutions or multinational corporations or even the state. Networks, he maintains, are the principal holders of power and counter-power in contemporary society but they do not act as single actors, instead they form strategic alliances with other actors. And the media is the space where these power struggles are played out – encompassing traditional mass media and, increasingly, the interactive horizontal networks based around the internet and wireless communication. Networks in this context are defined as: 'complex structures of communication constructed around a set of goals that simultaneously ensure unity of purpose and flexibility of execution by their adaptability to the operating environment. They are programmed and self-configurable at the same time' (2009: 21). Networks, according to Castells, became the most efficient type of organisation as a consequence of three of their principal features: *flexibility* (the ability to reconfigure in shifting contexts while holding onto their goals); *scalability* (growing or shrinking in size with little disruptive effects); and *survivability* (their ability to withstand attacks to their 'nodes' and 'codes' (2009: 23). He goes on to claim that:

> with the advent of nanotechnology and the convergence of microelectronics and biological processes and materials, the boundaries between human life and machine life are blurred, so that networks extend their interaction from our inner self to the whole realm of human activity, transcending barriers of time and space.
>
> (2009: 23–4)

These technologies have enabled them to co-ordinate functions on a global/local scale in real time. Information and the processing of information have become the organising principles of society and the basis of social action, affecting every sphere of human activity. Thus we live in a 'network society', as distinct from an 'information society' or 'knowledge society'.

A complex array of networks intersect at various points. Castells labels the controllers of these points of connection between different strategic networks 'switchers' (for example, science and technology networks, business networks and media networks) and the actors who programme/reprogramme the networks in terms of their goals, 'programmers' (2004: 32–3; see also 2009: 45). The switchers and programmers will vary depending upon the specific network. It is here, he claims, that power operates most decisively – not in the sense of a power elite wresting control but a more subtle process whereby framing and agendasetting take place within diverse, flexible communication systems that are continually in flux. Thus he argues:

> It is precisely because there is no unified power elite capable of keeping the programming and switching operations of all important networks under its control that more subtle, complex and negotiated systems of power enforcement must be established.
>
> (2009: 47)

According to Castells, 'power is the relational capacity to impose an actor's will over another actor's will on the basis of the structural capacity of domination embedded in the institutions of society' (2009: 44). Power is seen as differentiated and resistance is achieved through a variety of means, such as blocking the switches of connection between networks and reprogramming dominant codes within the networks' programmes. Environmental pressure groups campaigning around climate change, for example, exercise counter-power through disrupting switching and resisting programming. In sum, four principal types of power are identified as existing within the network society (see Castells, 2009: 42):

(1) *Networking power* – this refers to the key role of *gatekeeping* to include or exclude social actors or organisations from global networks. Those who control the operations of each communication network are the gatekeepers and they have power to block or allow access (Castells, 2009: 42).

(2) *Network power* – this is power that results from the imposition of rules of inclusion/standards or protocols of communication that need to govern social interaction within networks (see Castells, 2009: 43).

(3) *Networked power* – this refers to the editorial decision-making power of news organisations, their *agenda-setting* power. This includes decisions made by owners, managers, editors, journalists, industry, government and social elites which are underpinned by the overall goal to appeal to mass audiences and increase profit (see Castells, 2009: 419).

(4) *Network making power* – this is the capacity to set up and programme/reprogramme a network, and for Castells it is the most important form of power. Network making power is principally in the hands of media owners and controllers such as Rupert Murdoch or Silvio Berlusconi who are at the helm of global multimedia conglomerates that are closely bound up with business networks. However, most decisions can be seen as the outcome of complex interactions among networks with common vested interests; power does not reside so much with single actors but with strategic alliances between networks (see Castells, 2009: 42–7, 420).

The 'programmers' and 'switchers', due to their position in the social structure, are seen as possessing the most significant form of power within the network society – 'network making power' – defined as the capacity to set up and programme/reprogramme a network (Castells, 2009: 47, 418–20). Castells provides various empirical examples of how this occurs (mainly based on the United States) including the anti-globalisation movement, the environmental movement and the Barack Obama presidential campaign. He calls the new forms of networked communication 'mass self communication'; 'mass' because an individual posting (for example, a YouTube video or a blog) has the potential to be consumed by a global audience (Castells, 2009: 55). This is mass self-communication because the messages are self-generated by the individual and potentially reach thousands of people, and involves audiences interacting with the media in creative ways.

New spaces of flows are created whereby internet users increasingly rely on search engines to locate information across the globe rather than physically visiting a library or government office, for example. In the network society he contends that 'clock time' – the disruption of biological rhythms associated with Taylorism and industrial production – is replaced with 'timeless time' through, for example, the instantaneous processing of financial transactions. This is challenged by Lash and Urry's concept of 'glacial time' – a more cosmological perspective taken by environmental groups that views life in terms of the evolution of the planet (Lash & Urry, 1994). Castells argues that digital media was key for Stop Climate Chaos, a coalition of more than 70 NGOs that was formed in 2005. The internet helped to co-ordinate activists around the world and played an important role in their media strategy (2009: 324). Thus he claims: 'Internet-mediated social networks are key ingredients of the environmental movement in the global network society. The Internet has extraordinarily improved the campaigning ability of environmental groups and increased international collaboration' (2009: 325). The fluid nature of the internet is well suited to the loose, non-hierarchical character of many grassroots environmental movements. At the same time he rightly acknowledges that targeting digital outlets forms part of a wider multimedia strategy whereby mainstream media are frequently utilised by established NGOs, such as Greenpeace, alongside new modes of communication.

the versatility of digital communication networks has allowed environmental activists to evolve from their previous focus on attracting attention from the mainstream media to using different media channels depending on their messages and the interlocutors they wish to engage. *From its original emphasis on reaching out to a mass audience, the movement has shifted to stimulate mass citizen participation by making the very best of the interactive capacity offered by the Internet.*

(2009: 327 original emphasis)

Castells clearly sees mass self-communication as providing the potential for new social movements to challenge social attitudes and values, encouraging a greater range of voices to be heard but not determining the content or direction that resistance may take. However, at the same time he observes that it is also a space where power holders themselves are continually seeking to reassert their own dominance; a key part of which involves attempts to restrict the liberating potential of mass self-communication.

Gaining visual prominence is related to cultural capital. Castells rightly notes how celebrities have an increasingly enhanced presence through the web and they are clearly appearing as key voices in many debates about environmental issues (Boykoff & Goodman, 2009). Indeed, the very definition of what constitutes a celebrity or an opinion leader in western society has widened and we have seen the rise of micro-celebrity activists who are not celebrities in the usual sense of the word (see Anderson, 2011, 2013; Tufekci, 2013). Although Castells acknowledges that environmental groups increasingly involve celebrities in backing their causes (including stars such as Angelina Jolie, Brad Pitt and Leonardo DiCaprio), there is little discussion of the challenges and hurdles this may pose.

Celebrities may soften or mute the message and their involvement can, on occasion, backfire. They may have different agendas and their voices may drown out other actors in the movement (Tufekci, 2013). As Meyer and Gamson observe: 'in constructing their legitimacy to speak for a movement, celebrities frequently alter the claims of that movement to more consensual kinds of politics...the very spotlight of notoriety that comes with celebrity participation may drown out some movement claims and constituents' (1995: 181–7). Also, a recent study casts some doubt as to their ability to further awareness of environmental issues in the long term. The research sought to analyse the impact of different visual representations of climate change (including images of celebrities) drawn from a sample of national press coverage during 2010 in the US, UK and Australia (O'Neill et al., 2013). The findings suggest that the involvement of celebrities in climate change campaigns does little to increase issue salience – the extent to which people think that the issues are personally important. Images of well-known figures – including Bob Geldof and Richard Branson – tended to be viewed cynically and evoked fairly strong feelings that climate change was unimportant. Such imagery of prominent personalities appeared to have little impact on people's perceptions about whether they could do something about climate change in the UK and US. However, the Australian participants exhibited strong feelings of disempowerment when presented with pictures of politicians and celebrities, perhaps reflecting the highly politicised nature of the climate debate in Australia (see Anderson, 2013).

An IPSOS MORI survey of 1,000 people living in the UK aged between 16 and 64 commissioned by Climate Week found that multi-millionaire entrepreneur Sir Richard Branson (10 per cent) was viewed as the person most likely to make them act on climate change, followed by Bill Gates, business magnate and former chief executive of Microsoft (9 per cent), and Prince Charles (8 per cent) (IPSOS MORI, 2012). Richard Branson also came top among UK respondents in an internet survey of climate champions undertaken in 2007 (AC Nielsen, 2007). Slightly over half of the respondents (56 per cent) said that none of the personalities listed would make them act on climate change. The numbers of people saying that they would be persuaded by pop celebrities such as Arnold Schwarzenegger, Gwyneth Paltrow or Leonardo DiCaprio was for each individual a miniscule 1 per cent. To put this in context, when asked who would they trust the most if they were giving views on climate change, 66 per cent opted for 'scientists', a combined 24 per cent went for 'family/friends' and only 1 per cent selected 'celebrities'. Interestingly the list did not include environmental organisations as an option. On balance then it seems that celebrities play more of a mobilising role rather than bringing about a fundamental change in social attitudes (Thrall et al., 2008).

Castells sees global media groups, such as News Corporation, as key social actors since they programme media networks. Moreover, they exercise networking power over other actors, such as political actors and corporate businesses (2009: 422). Rupert Murdoch, media tycoon and founder of News Corporation, is seen as a key 'switcher'. He connects media, cultural, political and financial networks. He provides access and transfers resources among these networks, though his principal source of power is seen as residing in the media (2009: 429). However, these programmers and switchers are not individuals; they are network positions embodied by social actors – thus Castells points out that 'Murdoch is a node, albeit the key node, in one particular network: News Corporation and its ancillary networks in media and finance' (2009: 429).

As we shall see in Chapter 4, following Murdoch's lead, News Corporation outlets went green from the UK's *The Sun* newspaper to Fox International Channels and featured environmental awareness campaigns, though this was soon to fade as the economic recession began to bite (Brainard, 2006). Indeed, Murdoch donated US$500,000 in 2006 to former US President Bill Clinton's Global Climate Initiative.

Another increasingly prominent voice in environmental debates is that of Sir Richard Branson, Virgin CEO, who in the same year (2006) pledged to donate £1.6 billion (the equivalent of all the profits from Virgin Atlantic and Virgin Trains for the next ten years) to renewable energy initiatives and research in biofuels to reduce climate change. Richard Branson's son, Sam, travelled with his father and a film crew to the Arctic and published a diary about their adventures in 2007. In 2012 he was a key backer of the Greenpeace 'Save the Arctic' campaign mentioned earlier (see Anderson, 2013). Branson has placed social media at the heart of his business branding and at the time of writing he has 3.6 million followers on Twitter and 692,000 'likes' on his Facebook page. He regularly uses these means to send messages in support of environmental initiatives (see Figure 2.1).

Thus far, however, we have focused on the West. What about developing countries that are more reliant on traditional modes of communication? Is Castells' theory equally applicable to the peoples of more rural and remote regions that are less connected with these technologies?

Figure 2.1 Virgin Green Angel RT @richardbranson: It's clearer than ever: #climatechange is real, humans are the cause, and we have to act http://t.co/q7e3ucolZC, 14 October 2013

The digital divide

While Castells is right to point to the growing importance of digital technology in the developed world and their liberating potential, he pays insufficient attention to the digital divide. Take Nepal, for instance. It is situated in South Asia in the Himalayas and it is one of the poorest countries in the world. Nepal is considered to be the fourth most vulnerable country in the world to climate change and is particularly at risk from floods, droughts, sea-level rise and storms. It has a population of over 29 million people with approximately 80 per cent living in rural areas dependent on subsistence farming. Nepal is ranked 157 out of 186 countries on the Human Development Index and 70 per cent of the population live on less than £1.50 a week (the equivalent of $2). Far from this being a 'networked society', the majority of people receive their news from radio, and to a lesser extent television (see Table 2.1 below). Internet use is low (the penetration rate is around 9 per cent) and it is mainly accessed by young urban-dwellers (Colom & Pradhan, 2013).

Much of the internet use is concentrated in the more developed Kathmandu Valley region that is dotted with many cyber cafes, but connectivity tends to be slow and unreliable. Although mobile phone ownership is widespread and growing across the country, mobiles are generally just used for calls and texting, and by males rather than females (see Colom & Pradhan, 2013). Blogs were influential during the April 2006 uprising against the royal-military coup staged by King Gyanendra in February 2005, but these do not generally have a wide reach (Routledge, 2010).

The BBC Media Action project, Climate Asia, undertook a nationally representative survey of attitudes towards climate change and communication between May 2012 and March 2013 involving 2,354 households. They also undertook interviews with 20 opinion leaders and experts in Nepal, together with five community assessments and 12 focus group discussions across the country. The research found that the majority of rural communities used local community networks to share knowledge, such as women's groups. The greatest concern of the survey participants was not having enough food to eat and not having enough income to provide for their children's education. Approximately two-thirds (66 per cent) claimed that they were very worried about climate change impacts (especially the effects on agricultural productivity) and this was perceived as influenced both by personal experiences of extreme weather and by media reports. Of all the Asian countries surveyed they

Table 2.1 How people use the media in Nepal now

	All	Male	Female	Far-Western	Mid-Western	Western	Central	Eastern
Base: All respondents	2,354	1,212	1,142	195	315	542	826	476
%	%	%	%	%	%	%	%	%
Mobile phone	64	73	53	44	49	76	65	64
Radio	54	68	39	53	51	56	56	50
TV	46	49	44	17	19	68	43	59
Internet	6	8	3	1	1	12	5	4

Source: Colom and Pradhan (2013: 53).

were found to be the most willing to take action but they were struggling to adapt. A total of 59 per cent rated their media exposure as 'low' and 80 per cent mentioned lack of access to information as a major barrier. When asked how they would most like to be provided with information about changes in water, food and energy supplies their preferred source of information was radio (86 per cent) and then television (77 per cent). Only 8 per cent selected mobile phones (including SMS) and 6 per cent selected the medium of the internet (see Table 2.2 below).

With low levels of literacy in the very poor rural areas it is clear that these people do not consider themselves players in the 'network society'; with the exception of mainly young, male, comparatively wealthy urban-dwellers, they have few means of 'mass self-communication'.

While this may be a somewhat extreme example, digital skills are often low even in high access countries (Fuchs, 2009). Additionally, Christian Fuchs argues that if we look at the top 20 Web 2.0/Web 3.0

Table 2.2   Preferred sources of information on these issues

Source: Colom and Pradhan (2013: 53).

platforms, 16 of these are owned by organisations that are based in the US. This includes Facebook, YouTube, Blogger, Wikipedia, MySpace and Twitter. The only one of these 20 platforms that is non-profit and advertising-free is Wikipedia. Mass self-communication is already being co-opted into corporate frames (Couldry, 2012). For Fuchs users of Web 2.0/Web 3.0 constitute an 'exploited class of knowledge workers' since companies continually track their online activities and they are exposed to personalised advertising (2009: 95).

Attention and visibility

In addition to inequalities of access Castells downplays the fact that not all players are equally visible on the web. The top sites tend to be the online outlets of traditional mass media such as BBC Online and CNN Online. Alternative news sites such as Indymedia are ranked way down the list in terms of the numbers of people accessing the site; it sits at number 16,437 compared with BBC Online (ranked 53) and CNN Online (ranked 58). As Fuchs observes:

> a central filter of the Internet that benefits powerful actors is formed by visibility and the attention economy. Although everyone can pro-duce and diffuse information in principle easily with the help of the Internet because it is a global decentralized many-to-many and one-to-many communication system, not all information is visible to the same degree and gets the same attention. The problem in the cyberspace flood of information is how in this flowing informational ocean other users draw their attention to information.
>
> (2009: 96)

To be fair Castells does go some way to acknowledge this when he concedes:

> gatekeeping still yields considerable networking power because most socialized communication is still processed through the mass media, and the most popular information web sites are those of the main-stream media because of the importance of branding in the source of the message. Furthermore, government control over the Internet and the attempts of corporate business to enclose telecommunica-tions networks in their privately controlled 'walled gardens' show the persistence of networking power in the hands of the gatekeepers.
>
> (2009: 419)

As Nick Couldry (2012) observes, a key issue concerns whether Castells' theory can adequately explain political and social agency. What other institutional factors need to be present if media appearances are to divert the actions of interlocking networks of power? A case in point is the recent UK protests against a badger cull in 2013 which had the high profile celebrity support of Brian May, lead guitarist of the rock band Queen, and the backing of numerous scientists. The petition to Number 10 Downing Street attracted the largest ever number of signatures yet the cull still went ahead despite considerable media exposure, overwhelming public concern and widespread opposition among scientists. This included considerable coverage on the web through tweets, Facebook 'likes' and so on. In the following section we consider how far these have opened up new opportunities for environmentalists in bringing attention to a cause and the extent to which this impacts on policy.

The blogosphere, social networking sites and protest

With the cutbacks affecting journalism, bloggers are increasingly supplementing news coverage. Blogs are providing updates about what is happening in faraway locations such as Antarctica (Thorsen, 2009). It is becoming increasingly difficult to distinguish between blogs and mainstream news and many green blogs have become increasingly viewed as influential and authoritative sources (see Cox, 2013) – for example, climate scientist Michael Mann's blog. Many environmental reporters have developed their own blogs as well – most notable of which is Andrew Revkin's dot Earth blog at the *New York Times*. However, at the same time there are also numerous anti-environmentalist blogs that are critical of the science of climate change, for example.

Social networking sites are clearly becoming increasingly important for environmental campaign organisations. Protest action has become more personalised and social media have become a vehicle of self-presentation with the ability to share and 'like' content, and 'favourite' and 'retweet' posts. Facebook, which was launched in 2004, had 699 million daily active users on average in June 2013 (http://newsroom.fb.com/Key-Facts). It is the largest social networking site in Europe and the US (van Dijck, 2013). From April 2006 organisations were able to register on Facebook and now most large environmental NGOs have their own Facebook page. Location based services (such as Foursquare) give environmental groups the ability to document where they visit through using GPS (global positioning system). As we shall see in Chapter 5, this proved to be a useful tool in the BP oil disaster where Gowalla (a location based social network launched in 2007) enabled environmental activists to tag locations affected by the spill. All these technologies, however, can also make it easier for authorities to track individuals and undertake surveillance. Repressive regimes have also become increasingly sophisticated in controlling and filtering access to information (Morozov, 2012).

Nowhere is this perhaps more of an issue than in China. Social media have been an important means of co-ordinating environmental action in mainland China and environmental campaigners have embraced the opportunities provided by the internet (see Xie, 2011a, 2011b). In November 2009 over 1,000 people (mostly drawn from the middle-classes) took to the streets in Guangzhou, China, to protest against the building of a rubbish incinerator near their homes. Twitter was used as a means of sending real-time reports of the unfolding of events and photos were posted on the internet.

Twitter

Twitter is a 'micro-blogging' application that allows users to broadcast real-time short messages of 140 characters or under. The default setting is to make these messages public. Since it was launched in 2006, Twitter has become the largest micro-blogging site on the internet. At the time of writing in 2013 there are estimated to be 135,000 new Twitter users signing up a day around the globe but around 40 per cent of people just follow others and do not tweet. Also, there are a number of fake accounts. Currently North America has the highest number of Twitter users. A total of 16 per cent of internet users in the US are now active on Twitter reaching over 90 million US people monthly (https://www.quantcast.com/twitter.com).

Both Facebook and Twitter have been blocked in mainland China since the ethnic riots of 2009, which the government blamed on social media. Although the Chinese government censors Twitter, growing numbers of people are accessing restricted internet and social networking sites via a virtual private network (VPN) abroad. As the authorities clamp down individuals are becoming increasingly creative in finding ways around it. At the time of writing in 2013 there are reports that the Chinese government may be about to relax some of its strict censorship laws in the free-trade business district of Shanghai.

The Arab uprisings that spread across North Africa and the Middle East in early 2011 also led to numerous attempts by the authorities to censor the internet as well as control access to information by foreign journalists, including through personal intimidation (Gerbaudo, 2012). Clearly concerned about the evident power of social media to amplify dissent the Mubarak regime shut down all internet and mobile phone communication on 28 January 2011 for almost a week. However, this tactic had limited success since by that point face-to-face communication had become more important. In fact this move seems to have had the opposite effect to that intended; ironically this attempt to limit personal freedoms angered a great many people who were previously undecided and actually spurred them onto the streets (Gerbaudo, 2012).

This brings us to the much-debated question of whether Twitter and Facebook were instrumental in bringing about revolution and the implications of this for environmental protest. The Arab uprisings were not a single incidence highlighting the power of digitally networked media; the 2009 revolt in Iran and the Occupy movement in 2011 were also widely proclaimed as highlighting how new social media have enabled citizens to challenge traditional gatekeepers and news organisations. Platforms such as Twitter certainly aid co-ordination but to what extent do they trigger a fundamental challenge to the status quo? Journalist and best-selling author Malcolm Gladwell concludes in his article 'Why the revolution will not be tweeted' that this form of activism with its less personal and relatively weak ties does not generate long-lasting, truly effective collective action (Gladwell, 2010). He compares it with the strong ties and centralised, hierarchical structure of the civil rights movement of the 1960s. Social networking sites increase participation he concedes, but they do not increase motivation in the same way as earlier forms of activism, nor can they involve strategic planning:

> Unlike hierarchies, with their rules and procedures, networks aren't controlled by a single central authority. Decisions are made through consensus, and the ties that bind people to the group are loose...They can't think strategically; they are chronically prone to conflict and error. How do you make difficult choices about tactics or strategy or philosophical direction when everyone has an equal say?
>
> (Gladwell, 2010: 4)

Following the unfolding events of the Arab Spring uprisings this techno-pessimistic perspective has come under considerable attack. Nevertheless some very sweeping claims have been made about the power of

Twitter and other social networking sites to instigate social change so in this sense Gladwell is right to call for some caution. Clay Shirky, author of *Here Comes Everybody and Cognitive Surplus: Creativity and Generosity in a Connected Age*, provides one of the most techno-optimistic accounts of the new internet age. As information increases he sees the possibilities of collective dissent growing by the minute:

> as the communication landscape gets denser, more complex, more participatory, the networked population is gaining greater access to information, more opportunities to engage in public speech, and an enhanced opportunity to undertake collective action. In the political arena, as the protests in Manila demonstrated, these increased freedoms can help loosely coordinated public's demand change.
>
> (Shirky, 2011)

But does clicking on links, 'liking' comments and signing online petitions really constitute a deep engagement with the issues?

Information overload and clicktivism

'Clicktivism', sometimes referred to as 'slacktivism', constitutes a relatively passive way of interacting with the media and may encourage the false impression that one is really making a difference merely by clicking on a link. Willard, for example, argues:

> At a time in which we need tools to be able to communicate the complexity and integratedness of issues, mainstream SNSs seem to be leading to 'dumbing down' sustainability to a lowest common denominator. If people begin to believe that they are 'doing their part' by sending virtual fish to each other, we are in trouble. Even more dangerous is the likelihood that sustainable development values will simply be overwhelmed by the relentless drive towards nostalgia and consumerism embodied in mainstream SNSs.
>
> (2009: 27)

For this reason many environmental advocacy groups prefer to contact their supporters via email and Facebook is considered very much as a secondary option (see Cox, 2013). In response to arguments about 'slacktivism' Shirky argues: 'the fact that barely committed actors cannot click their way to a better world does not mean that committed actors cannot use social media effectively.' And in response to the other criticism that governments are getting much better at monitoring, interdicting or co-opting these tools he claims:

> Indeed, the best practical reason to think that social media can help bring political change is that both dissidents and governments think they can. All over the world, activists believe in the utility of these tools and take steps to use them accordingly. And the governments they contend with think social media tools are powerful, too, and are willing to harass, arrest, exile, or kill users in response. (2011)

There are two potential issues here: first there is the tendency to isolate social media such as Twitter from the wider technological and social contexts in which it operates and second there is a risk that we simply assume their defining political features rather than base this on empirical research (see Segerberg & Bennett, 2011). Rather than seeing the offline and online worlds as separate it is more accurate to view them as intermeshed. As Segerberg and Bennett observe: 'Digitally networked protest spaces often involve dense webs of technologies deployed by different actors, so Twitter is one of potentially many digital mechanisms that co-constitute and co-configure the protest space' (2011: 201). It should be remembered that not all Twitter users are as influential as others and it assigns different weight to different voices. Celebrities and politicians tend to have the most frequently followed accounts that are verified and they employ PR staff to manage them. As van Dijck observes:

> The ideal of an open and free twitterverse in reality comes closer to a public dialogue ruled by a small number of hyper-connected influencers skilled at handling large numbers of followers. The platform's architecture privileges certain influential users who can increase tweet volume, and who thus garner more followers.
>
> (2013: 74)

Also, although ordinary citizens can exert influence through, for example, 'liking' and 'trending' and are able to tap into crowd-sourcing there is a world of difference between this and the kind of power that owners can exert: 'Although both owners and users can manipulate social media's filtering apparatus, it is important to distinguish their difference in power' (van Dijck, 2013: 158).

It is worth bearing in mind that while many were quick to label the uprisings in Egypt as 'Twitter Revolutions' or 'Facebook Revolutions' only a small proportion of local people were directly mobilised via such means since, just as we saw with Nepal, internet penetration rates were low. Only around a quarter of households in Egypt were connected to the internet in 2011, only 4 per cent of adults used Facebook and a tiny 0.15 per cent had their own Twitter account (see Gerbaudo, 2012: 49). Research undertaken by the Tahir media project found that only a small number of those who took part in the uprisings appeared to have been motivated by social media. Interviews with protestors in Cairo revealed that only 16 per cent used Twitter and 42 per cent used Facebook. Word of mouth and traditional media channels were the main means of mobilisation for the majority of those who took to the streets (Gerbaudo, 2012).

Both the techno-optimistic and techno-pessimistic arguments are equally flawed. As Cottle rightly observes, claims that deny the role of social media as a potent tool for bringing together diverse transnational networks and channelling protest in new and creative ways underestimate its potential. Furthermore exaggerated assertions about the power of social media to bring about social change tend to cast the media in a narrow technologically deterministic lens:

> Inflated claims about the power of new social media to foment protest and revolution lend themselves to the charge of media centrism and technological determinism that obfuscate the preceding social and political forces at work as well as the purposive actions of human beings prepared to confront state intimidation and violence to bring about political change.
>
> (Cottle, 2011: 298)

There is no doubt that both mainstream media and new modes of communication potentially play an important role in environmental protest but it is worth remembering that digital media are just one element in a strategic toolbox. Moreover, as Castells (2007) acknowledges, political uses of the internet are only sometimes about bypassing 'old media'. As Gavin rightly notes: 'Web mobilisation cannot be assessed in isolation from other contextual factors, and Web active groups are still subject to a range of imperatives and limitations only tangentially related to their web presence' (2009: 140).

There is considerable variation among environmental groups – some are more formal and hierarchical than others, whereas others are more informal with loose, fluid formations. Also, they shift over time and employ different organisational repertoires at different points.

Environmental groups face a number of constraints in reaching a wider audience when there is increasing information saturation and a very crowded media. The flood of content means there is greater competition for attention and the fragmentation of audiences is likely to increase even more (Cox, 2013).

Indeed, the most important public debates now take place via 'public screens' whether this be television, computers or hand-held mobile devices (see DeLuca & Peeples, 2002). Here politicians and global corporations stage spectacles and activists perform what DeLuca calls 'image events' which seek to bring the former to account (DeLuca, 1999). New forms of technology promote new modes of perception. As DeLuca et al. note:

> With the spread of smartphones, space and time cease to be barriers to living in a mediated world all the time. We need no longer go to a medium or find an Internet connection, for they are in our pocket, a part of us. When people widely adopt and deploy in expected and unexpected ways a new medium such as smartphones, they transform a host of practices and contexts, including activist, business, consumer, interpersonal, journalistic, leisure, organizational, parenting, and pedagogical practices.
>
> (2012: 486)

Image events themselves are certainly not new. Take, for example, Greenpeace's early direct action against Soviet whaling in the mid-1970s, which did not save the whale but received worldwide television coverage and, in the words of its then Director, Robert Hunter, appeared to explode 'in the public's consciousness to transform the way people view their world' (cited in DeLuca & Peeples, 2002: 136). As Paul Watson, now at the helm of Sea Shepherd Conservation Society, explains:

> When we set up Greenpeace it was because we wanted a small group of action-oriented people who could get into the field and, using these McLuhanist principles (for attracting media attention), make an issue controversial and publicize it and get to the root of the problem.
>
> (quoted in Scarce, 1990: 101)

Alongside the television images of activists in inflatable rubber dinghies pitting themselves between the harpoons and the whales, Greenpeace launched petitions asking national governments to apply international pressure and distributed flyers to drum up more public support (Greenpeace, 2011). Policy change can take many years to occur and is dependent upon a number of different factors. Around a decade after the Greenpeace anti-whaling action in 1975, the International Whaling Commission (IWC) placed an indefinite moratorium on commercial whaling. This moratorium came into effect in 1986 and was in large part due to countries that did not participate in commercial whaling, but were concerned about the plight of the whale, joining the commission in greater and greater numbers. Indeed, between 1973 and 1982 membership of the IWC increased from 14 member nations to 39 (Hoel, 1986). Currently there are 89 member countries. Japan, Norway and Iceland have managed to find loopholes in the moratorium which has allowed them to continue commercial whaling either through invoking the scientific research clause or their right to opt out. To a large extent, then, the campaign has been successful but there still remain significant regulatory and cultural hurdles.

Image events created by environmental activists today, such as those employed in the Save the Arctic campaign, increasingly encompass a mix of parody, satire and irony as highlighted earlier in this chapter. They involve the remixing of old and new, exposing the fake quality of much news itself. This way of critiquing news has a long history and may be traced back to détournement and the French Situationist International movement of the 1960s (see Russell, 2011). The culture jamming group, The Yes Men, are masters of this fakery recognising the potential power of comedy and satire to draw attention to global injustices and corporate spin. Since the late 1990s they have posed as numerous officials from the world of business and politics and worked with a variety of NGOs on their media campaigns. As Russell observes:

> Fake news and other forms of political remix both within and outside mainstream outlets represent key components to networked journalism. Not only do they influence news discourse, they signal a larger phenomenon: the evolution of journalistic forms and the emergence of a space for dissent being carved out in the middle of the spectacle.
>
> (2011: 126)

The contemporary media scene is thus characterised by what DeLuca labels 'panmediation' whereby in most societies a multitude of media coexist and interact (DeLuca et al., 2012: 487). User-generated content feeds into offline media and vice versa, which creates a complex dynamic media ecology of overlapping flows. The key question, as we shall explore in subsequent chapters, is whether short-term symbolic gains obtained by environmental groups translate into long-lasting legitimacy.

Summary

I began this chapter by illustrating what I see as a new spectacularisation of the environment in late modern society by examining the strategies employed by Greenpeace in its Save the Arctic campaign. Environmental groups have long deployed resonant symbols, but the contest for power is more visibly played out on the media stage than at any time before. Any attempt to characterise society as either 'risk society' or 'network society' inevitably produces a rather broad-brush account that, although insightful, cannot adequately account for the complexities and contingencies of power. Beck tends to present us with a static, a-historical account. By contrast Castells' account is more dynamic since he recognises that networks are continually reconfiguring around new issues and changing coalitions. Yet both pay insufficient attention to back-stage, less visible aspects of power and it is to this that the discussion now turns in the following chapter.

Further reading

Allan, S. (2013) *Citizen Witnessing: Revisioning Journalism in Times of Crisis.* Cambridge: Polity.

Beck, U. (2009) *World at Risk.* Cambridge: Polity Press.

Castells, M. (2009) *Communication Power.* Cambridge: Polity. Chapter 5.

Cox, R. (2013) *Environmental Communication and the Public Sphere* (3rd Edition). London: Sage. Chapter 7.

Gerbaudo, P. (2012) *Tweets and the Streets: Social Media and Contemporary Activism.* London: Pluto.

Hassan, R. (2004) *Media, Politics and the Network Society.* Maidenhead, Berkshire: OpenUP.

Howard, P. (2011) *Castells and the Media.* Cambridge: Polity.

Mythen, G. (2004) *Ulrich Beck: A Critical Introduction to the Risk Society.* London: Pluto Press.

Russell, A. (2011) *Networked: A Contemporary History of News in Transition.* Cambridge: Polity.

3
News Agendas, Framing Contests and Power

The great blockbuster myth of modern journalism is objectivity, the idea that a good journalist or broadcaster simply collects and reproduces the objective truth. It has never happened and never will happen because it never can happen. Reality exists objectively but any attempt to record the truth about it always and everywhere necessarily involves selection.

(Nick Davies, 2008: 111)

All I know is just what I read in the newspapers.

(Will Rogers, American political humourist)

What gets in the news and how news is portrayed is a product of struggles between groups who hold competing definitions of reality. In earlier times newspapers were often one of the main means by which people in Western societies received information about issues of public importance. Now of course we have a multitude of media platforms including television and smartphones and we can upload wikis, blogs, tweets and YouTube clips at the touch of a button. How influential is this increasingly fragmented media scene on our cognitive maps of the world and, in particular, our attitudes towards environmental issues? And what determines the opportunities of news sources to shape news agendas?

This chapter begins by examining the concepts of 'agenda-setting', 'agenda-building' and 'framing' in order to explore the selective nature of environmental coverage. Environmental issues may be framed in a number of different ways depending upon the influence placed upon such factors as economic development or biodiversity. Different stakeholders have varying degrees of power in commanding news media attention and being treated as credible news sources. The chapter examines whose views tend to gain visibility and considers the changing strategies of NGOs in relation to traditional news media and online news.

Agenda-setting

Why do some social problems become defined as issues and others not? It is this key question that the agenda-setting model seeks to address. It starts from the premise that if the media report frequently on an issue and this receives prominence then audiences will see it as more important. As Bernard Cohen famously stated: 'The press may not be successful much of the time in telling people what to think, but it is stunningly successful in telling its readers what to think about' (Cohen, 1963: 13).

The agenda-setting process involves three main elements: the media agenda, the public agenda and the policy agenda (see Figure 3.1 below). The agenda-setting model asserts that the news media may not necessarily tell us *what* to think, but they are strikingly influential in terms of telling us what to think *about*. In other words, the presence, absence or positioning of a news item (e.g. whether it is a lead story or relegated lower down in terms of the hierarchy of stories) can impact on their rankings on public and policy agendas. There is often a remarkable consensus among news organisations over the leading news stories of the day. In recent years the economic recession, for example, has dominated the headlines. According to this highly influential perspective, the

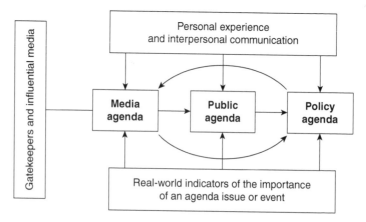

Figure 3.1   Three main components of the agenda-setting process
Source: Rogers and Dearing (1988) in McQuail, D. and Windahl, S. (1993) *Communication Models for the Study of Mass Communications*. Essex: Longman.

media are seen as influencing the public's priorities far more than pu concerns influencing the media.

A related concept is that of 'priming' which refers to: 'the process in which the media attend to some issues and not others and thereby alter the standards by which people evaluate objects in the real world' (Severin & Tankard, 2001: 226). It goes one step beyond the agenda-setting theory by also examining the role of the news media in influencing the audience's evaluation of information. Also relevant is the concept of 'gatekeeping' which refers to: 'the process of selecting, writing, editing, positioning, scheduling, repeating and otherwise massaging information to become news' (Shoemaker et al., 2009). Simply put, it is the ability of news workers to decide what gets in the news and what is left out and is influenced by journalistic conventions and computer algorithms.

Many researchers have explored the relationship between what publics say they are most concerned about and attempted to trace how far this corresponds with the hierarchy of issues portrayed by the news media. The findings on the influence of agenda-setting for environmental news have tended to be rather inconsistent, with some studies finding strong effects and others concluding that there is little evidence of a significant relationship (see Anderson, 1997; Cox, 2013).

The ability to keep issues off the agenda

What is particularly interesting, and much less studied, is the ability of news sources to keep issues *off* the news media agenda. Indeed, control over the media is as much to keep certain issues marginalised or hidden, as it is to publicise them (Anderson, 2009). While early agenda-setting theories rather simplistically assumed that there was a neat causal relationship between the hierarchy of issues in media agendas and that of public agendas, the more recent emphasis on agenda-building has moved the field significantly forward. Agenda-building refers to how different groups seek to influence what issues journalists report on and how they cover them (Cobb & Elder, 1972). It is now recognised that there may be a number of intervening factors that can affect agenda-setting, such as the amount of interpersonal discussion of the topic and the intensity of the initial coverage.

The end of agenda-setting?

With the rise of electronic media there are many more agendas in circulation and this has led some to propose the end of agenda-setting, as audiences fragment and content becomes more individualised

(McCombs, 2005). Given the sheer amount of information that we can access 'narrowcasting', the targeting of media messages towards specific segments of the public, has become more important. This clearly alters the pattern of news consumption and provides a different system of gatekeeping and filtering (Bennett & Iyengar, 2008, 2010). The explosion of social media also makes it more difficult for news sources to keep issues off the agenda. Given that it is now much easier for news organisations to find out what their audiences are interested in, could it be that reverse agenda-setting may be more likely to occur where the public increasingly set the media agenda? Or is there a greater degree of interaction between media and public agendas? Indeed, Cox observes: 'With broad access and interconnectivity of Internet sites, scholars will need to rethink who – if anyone – controls access and what determines newsworthiness' (2006: 197). In their analysis of what they see as the weakening of political media effects Bennett and Iyengar conclude that while agenda-setting and priming may remain influential:

> As media audiences devolve into smaller, like-minded subsets of the electorate, it becomes less likely that media messages will do anything other than reinforce prior predispositions... An exception to this pattern may occur for relatively inattentive and politically nonpartisan citizens exposed to big stories that are repeatedly in the news, receive prominent placement, and echo throughout the multiple media channels from television, to radio talk shows, to blogs and email forwarding. Less saturated news topics may have little effect on opinion (even for attentive partisans) than strategically targeted messages by interest groups and online organizations such as moveon.
>
> (2008: 724–5)

In other words they suggest that as audiences fragment they will tend to largely encounter information that reinforces their prior views and are unlikely to encounter material that challenges them. However, as Holbert et al. (2010) argue it could just be that the sources of these effects are different. The notion of the end of agenda-setting presupposes that (a) large numbers of people have access to the internet (b) that they frequently access it and (c) that they scatter across a diverse range of online platforms. While there is plenty of evidence in Western society supporting the first two contentions, with more and more people accessing the internet and on a more frequent basis (see Kohut, 2013; PEW, 2012), there is little evidence to confirm the third contention as audiences tend to be highly concentrated (McCombs, 2005).

There also seems to be a complex inter-media agenda-setting process at work whereby social media and traditional media are dynamically interacting so that sometimes it is the former setting the agenda and sometimes the latter (Coleman & McCombs, 2007; Ragas et al., 2014). Research suggests that the blogosphere relies heavily on traditional media as sources and journalists often use blogs as a source of story ideas (see Messner & Watson Distaso, 2008; Tran, 2013). There is some evidence to suggest that exposure to the internet produces weaker agenda-setting effects compared with traditional media (Conway & Patterson, 2008). In sum, there is little evidence that we have entered into an era of minimal news media agenda-setting effects as some have suggested (Shehata & Stromback, 2013).

Many have argued that the concept of media framing extends the idea of agenda-setting by suggesting that the media do more than simply increase the salience of an issue but also increase the salience of particular aspects of that issue. It is to this that we now turn.

The concept of framing

In order to grasp complex ideas we often use frames in everyday life to organise ideas and make sense of issues. The concept of frame originates from the social constructionist tradition that emphasises the socially constructed nature of what we take as reality; facts do not simply 'speak for themselves'. Erving Goffman (1974) uses the notion of frame to draw attention to the way in which we make sense of social situations through particular schemata of interpretation of which we are often largely unaware. He suggests that as we cannot fully comprehend the complexity of the world we rely upon these 'primary frameworks' to reduce/handle information and help us interpret and label events. As George Lakoff observes, frames underpin all knowledge and evoke emotional responses:

> All of our knowledge makes use of frames, and every word is defined through the frames it neurally activates. All thinking and talking involves 'framing.' And since frames come in systems, a single word typically activates not only its defining frame, but also much of the system its defining frame is in. Moreover, many frame-circuits have direct connections to the emotional regions of the brain.
>
> (Lakoff, 2010: 71–2)

Frames should not be confused with specific policy positions as individuals may have different opinions on an issue but share the same overarching interpretative frame. For example, an ethics frame for biotechnology could be used in a positive light as emphasising the moral duty to prevent world hunger or in a negative light as 'playing God with nature' (Nisbet & Huge, 2006: 10).

Thus we may distinguish between individual frames and media frames (Scheufele, 1999). Entman, for example, talks about individual frames as 'information processing schemata' while the media frames are 'attributes of the news itself' (1991: 7). A media frame is 'a central organizing idea or story line that provides meaning to an unfolding strip of events' (Gamson & Modigliani, 1987: 143). Journalists use narrative structures to convey a story's meaning rather than simply present 'the facts'. How environmental issues are framed by the news media is of great significance since this can potentially influence what is viewed as legitimate and 'common sense'. As Gamson puts it: 'News frames make the world look natural. They determine what is selected, what is excluded, what is emphasized. In short, news provides a packaged world' (1985: 618).

Widely used in a number of different social science disciplines spanning sociology, psychology, literary studies and politics – to name but a few – the concept of framing thus refers to processes of selection and emphasis in news media production, which makes some aspects of a news story more salient than others (Gitlin, 1980; Priest, 1994; Miller & Riechert, 2000). The concept of news framing refers to the 'principles of selection, emphasis, and presentation composed of little tacit theories about what exists, what happens, and what matters' (Gitlin, 1980: 6). News values such as drama, novelty, controversy and geographical and cultural proximity, strongly influence selection processes for environmental news (see Anderson, 1997). As Robert Entman observes:

> Framing essentially involves selection and salience. To frame is to *select some aspects of perceived reality and make them more salient in the communicating text, in such a way as to promote a particular problem definition, causal interpretation, moral evaluation and/or treatment recommendation* for the item described. Frames, then, *define* problems – determine what a causal agent is doing and costs and benefits, usually measured in terms of cultural values; *diagnose* causes – identify the forces creating the problem; *make moral judgments* – evaluate causal agents and their effects; and *suggest remedies* – offer and justify treatments for the problem and predict their likely effects.
>
> (Entman, 1993: 55, original emphasis)

According to Entman's definition above, then, a frame consists of the following elements: (1) problem definition (2) diagnosis of the causes (3) moral evaluation of those actors/agencies attributed as causing the problem and (4) recommendations for treatment. For Entman it is the selection and salience of certain aspects of an issue rather than the issue itself and this distinguishes it from agenda-setting or gate-keeping (Scheufele, 1999; 107). The formation of frames comes from a number of influences including news media organisational routines, journalistic frames and frames derived from news sources (Anderson, 1997; Scheufele, 1999).

Through choice of words, imagery, emphasis and the inclusion or exclusion of information, news media present a particular view of the world and legitimise certain truth-claims over others. The signifying aspects of a news item provide a 'cognitive window' on the world (Pan & Kosicki, 1993: 59). Journalists use frames to develop interesting pieces that convey a story in meaningful ways through the use of metaphors and catchphrases, and making connections between particular events and issues (Entman, 2004). For example, Anders Hansen examined the national UK press framing of Greenpeace protestors campaigning against Shell's proposal to dump a redundant oil installation in the North Sea in 1995 (Hansen, 2000). While the left-leaning *Daily Mirror* framed Greenpeace campaigners in an overwhelmingly positive light – being referred to as 'heroic', 'daring' and 'idealistic' – the right-leaning *Daily Telegraph* described them using such terms as 'arrogant', 'militant', 'bearded' and 'rebels'.

Many scholars have observed how social movements tend to be portrayed in terms of the protest paradigm. That is, they tend to be framed negatively and as deviant, and often it takes a violent incident to trigger news media attention in the first place (see Atton, 2013; McCurdy, 2012). In the next section we turn to consider the power differentials in the relationship between challengers, dominant sources and the news media.

News sources and power

Framing thus involves selecting certain truth-claims over others and, in the process, denying or silencing rival versions of reality. Frames are shaped by claims-makers including politicians, scientists and NGOs; they do not occur in a political vacuum (Olausson, 2009). Social and political power is integral but it has often been neglected in the framing literature, as many prior studies tended to analyse it almost exclusively in terms of audience effects. As Carragee and Roefs observe: 'frames, as imprints of power, are central to the production of hegemonic meanings' (2004: 222). It is essential to consider the framing contests both internal and external to the news media (see Allan et al. 2010; Anderson, 1997). Claims-makers, just like journalists, are selective in how they package an issue through emphasising particular definitions or interpretations over others. Frames are the product of a complex range of factors, are culturally specific and shift over time, reflecting changes in the relationships between stakeholders (Iyengar, 1991). Indeed, framing may be seen as a kind of second-order agenda-setting (McCombs et al., 1997).

The concept of 'information subsidies' developed by Gandy (1982) draws attention to the increasing professionalisation of news sources and the development of PR material that fits journalists' criteria for newsworthiness. He defines it as: 'an attempt to produce influence over the action of others by controlling their access to and use of information relevant to these actions' (1982: 61). According to Gandy, government sources have the necessary economic base to allow them to package events in media friendly ways and are therefore likely to become routine sources. Over the last 30 years pressure groups and industries have devoted more and more resources into their media strategies, recognising that material has to be packaged in a particular way to draw news media attention (Anderson, 1991). As Reese argues: 'plugging in framing as just one more content element, against which to measure effects, risks continuing to ignore basic power dimensions' (2001: 9).

Journalist's frames are shaped by the frames that news sources employ and research suggests that generally speaking government sources, industry and scientists tend to gain advantaged access to the news media (Hall et al., 2013; Anderson, 1997; Manning, 2001). Stuart Hall's influential concept of 'primary definers' still has some relevance for understanding the relationship between news sources and the media today, despite undergoing extensive qualifications (see Anderson, 1997; Schlesinger, 1990). Hall et al. (2013) argue that official sources or 'primary definers' (such as government ministers and the courts) gain advantaged access to the media by virtue of their status as representative of 'the people', their standing in society or their claims to expertise and authoritative knowledge. The media are seen as 'secondary definers' through their role in transmitting the views of the powerful. Once an issue is framed in the formative stages of a debate by the news media it can be very hard to shift it to another interpretation (Hall et al., 2013; Nisbet et al., 2003). Writing in the late 1970s Hall and colleagues observed that there was: 'a systematically structured over-accessing to the media of those in powerful and privileged institutional positions...The result of this structured preference is that these "spokesmen" become what we call *primary definers* of topics' (1978: 58, original emphasis).

Similarly in an influential study of media coverage of the anti-war movement in the late 1960s Todd Gitlin remarked: 'Simply by doing their jobs, journalists tend to serve the political and economic elite definitions of society' (1980: 12). Journalists spend a great deal of time cultivating trust in their relationship with news sources and factors such as professional status, seniority, sector in society and institutional affiliation are taken into account when weighing up source credibility (Hansen, 2010; Reich, 2013). There is certainly plenty of research evidence to support the theory that journalists tend to rely on official sources (see Carlson, 2009). Indeed, this tends to be mutually beneficial for both parties; the news media gain authority through having used an official source and at the same time the news source's status as a reliable source of information is potentially enhanced. Also, the tendency for journalists to work as a 'pack' on particular news beats (for example, the environment beat) may further reinforce the tendency to over-access particular sources that are regarded as trustworthy (Carlson, 2009). Indeed, some have observed a growing practice for some news outlets to lift unattributed material directly from other media outlets (Phillips, 2013).

In a more recent critique of UK journalism investigative journalist, Nick Davies, argues that news selection is governed by ten rules (see below). For Davies this amounts to what he sees as 'a global collapse of information-gathering and truth-telling' (2008: 154). In effect this situation amounts to a form of primary defining and according to Davies it has arisen as a direct result of 'being taken over by a new generation of corporate owners' in the Western world who have slashed budgets, cut staffing levels and increased output to maximise profit (Davies, 2008: 153). He cites figures to suggest that there are now more individuals employed in PR in Britain (47,800) than there are journalists (45,000). He sums up his argument in this way:

> the rules of production of the news factory themselves impose their own demands as media outlets pick easy stories with safe facts and safe ideas, clustering around official sources for protection, reducing everything they touch to simplicity without understanding, recycling consensus facts and ideas regardless of their validity because that is

what the punters expect, joining any passing moral panic, obsessively covering the same stories as their competitors. Arbitrary, unreliable and conservative. Most worrying, however, this flow of falsehood and distortion through the news factory is clearly being manipulated, by the overt world of PR and the covert world of intelligence and strategic communications.

(2008: 255)

While these ten rules overlap somewhat, are rather sweeping and do not acknowledge variations among news formats, they do point to some general underlying trends. Today's global 24/7 news culture is characterised by round-the-clock deadlines, increased competition and

---

(1) **Rule One: Run cheap stories.** Publish stories that are safe and quick to publish. Slowly unfolding, complex stories that possess a high degree of uncertainty tend to be expensive to cover.

(2) **Rule Two: Select safe facts.** Favour factual information that can be attributed to official sources. Journalists are more likely to go with official interpretations as this lays them less open to attack.

(3) **Rule Three: Avoid the electric fence.** Keep away from facts that are dangerous to report. Official sources are preferred and reporters tend to defer anyone who can potentially cause damage to them or their media outlet.

(4) **Rule Four: Select safe ideas.** Toe the line. Reporters play safe by embedding facts within moral and political frameworks that do not challenge the status quo.

(5) **Rule Five: Always give both sides of the story.** Appear impartial. Balancing opinions provides a safety net for reporters.

(6) **Rule Six: Give them what they want.** The bottom line is stories must increase readership/audience ratings.

(7) **Rule Seven: The bias against truth.** Follow the news factory and cut out context and meaning. The profit drive leads news to become distorted since the pressure is for shorter and shorter items that favour certainty, simplicity and human interest.

(8) **Rule Eight: Give them what they want to believe in.** Feed news consumers with what they want, regardless of editorial judgement about whether this is right or wrong. The most profitable ideas are those that get reported.

(9) **Rule Nine: Go with the moral panic.** Tap into public values. This rule just applies at times of perceived crisis. Coverage that taps into the public mood fits commercial imperatives.

(10) **Rule Ten: Ninja Turtle syndrome.** Follow the crowd. The pressures to follow the pack and cover what is being widely reported on by other news outlets, even if the credibility of such stories is questionable.

Figure 3.2   Flat Earth news: Playing it safe
Source: Adapted from Davies, N. (2008: 114–154).

increasingly 'multi-skilled' journalists who have to work more flexibly than ever before by the increased casualisation of the workforce as well as digital convergence (see Cox, 2013; Mitchelstein & Boczkowski, 2009; PEW, 2013). The claim that staff numbers on national UK newspapers have been cut wholesale needs to be viewed with some caution. However, there is clear evidence that budgets for science and investigative reporting have been substantially cut and workloads significantly increased (Boykoff & Yulsman, 2013). These economic and organisational factors result in daily working routines becoming more pressured with little time to verify information face-to-face, leading journalists to rely much more upon pre-packaged material from news sources (Lewis et al., 2008). The job of the journalist thus becomes more about sifting through the mass of information that lands on their desk each day (via emails, blogs, Twitter feeds, press releases, wire services, electronic bulletin boards, etc.) and deciding what merits coverage, rather than actively searching for news. Citizen journalism is becoming more commonplace, weakening traditional journalistic conventions and blurring the lines as to what constitutes 'news'. The digital revolution has certainly enabled many challenger news sources to gain more power in framing their messages and greater visibility, but how much do they really control this process?

A tango or a tug-of-war?

The relationship between news sources and the media has been described both as a dance (where news sources generally take the lead) or a more competitive tug-of-war (Castells, 2004; Gans, 1979). Are they allies or foes? Is their relationship symbiotic or inherently conflict-ridden? The answer is far from clear-cut since the relationship between news sources and the media is dynamic and constantly shifting. Relations between media and environmentalists is characterised by a fundamental power imbalance; journalists are the ones that have the upper-hand. Nevertheless, the theory of primary definers, discussed above, is too static and reductive to be able to accommodate the complexities of source-media relations in a variety of changing contexts. What Carlson labels as the 'competitive definers' approach instead views news sources as engaged in a constant battle with one another to define news frames but it should not be equated with a pluralist view of the claims-making process. According to the 'competitive definers' perspective social dominance by itself is not enough to guarantee successful access to the news media (Anderson, 1997; Cottle, 2010). Such an approach necessitates going beyond merely examining media coverage to deduce source influence. Ethnographic studies of news production can offer rich insights into the factors that enhance or constrain news sources that are competing with one another to gain news entry. While it is not uncommon for researchers to acknowledge that news media frames are partially shaped by the activities of various stakeholders, often studies tend to adopt a media-centric bias through taking an internalist approach (see Schlesinger, 1990).

For example, even though Olausson observes that: 'Recognizing the power of and struggle between various stakeholders and their influence on the process of framing a certain issue is vital and is a central component of the analysis of frames in their totality' (2009: 3), her analysis is restricted to an examination of newspaper portrayals of climate change. In an otherwise insightful analysis there is no examination of news sources' strategic actions – this is simply read off from the coverage from which she concludes 'no framing contest' is going on (2009: 12). Other studies are restricted to examining journalists' accounts of how they deal with sources or sometimes there is a combination of textual analysis alongside interviews with journalists. As Schlesinger notes:

> *Internalists* produce their analyses of source behaviour either by interpreting what sources do by a reading of media content or by deriving conclusions from accounts given by practitioners of journalism of their interaction with their sources, or by combining both. There is nothing wrong with this, but it has its shortcomings.
>
> (1990: 72)

One of the key limitations is that such an approach is unable to reveal less visible aspects of news production processes and the hidden faces of power. Moreover, it is important to recognise the complexity of source-media relations, given the frequent non-attribution of news sources within news articles. If a news source has little visibility within the news media it cannot necessarily be assumed that there is no framing contest occurring. Pressure groups are sometimes used as unpaid researchers by journalists and their input is not attributed, or a news source may deliberately hold back from proactively targeting the news media at a particular point in time, preferring to concentrate upon seeking to influence policy arenas (Anderson, 2007). As Libby Lester puts it:

> The practices of news making, the strategic activities of sources, the symbolism that flows in and around news and the active involvement of news audiences clearly point to the fact that journalism does not

exist in an impenetrable bubble, but rather within a complex web of interactions, meaning making and power relations, many of which can only be revealed via 'behind-the-scenes' studies of journalistic and source activity.

(2010: 77)

In sum, then, the contest to gain favourable media coverage is not a level playing field since official sources tend to have greater financial resources and stocks of cultural capital. However, numerous studies have shown that while socially dominant sources tend to dominate risk coverage, particularly where a major crisis is concerned (see Anderson, 1997), they are not automatically guaranteed successful news entry. News media access is dependent on numerous contingencies that are both internal and external to the media. Non-elite sources, despite being resource poor may, on occasion, develop highly successful media strategies (Anderson, 1997). As Lester notes: 'This acknowledgement that, despite an uneven playing field, challengers are able to at least join the game and impact on framing of an issue is vital and helps explain the broad range of conditions that influence and shape media coverage of the environment' (2010: 52).

However, gaining access and achieving coverage is only half the battle. How news sources claims are framed, and whether they are given legitimacy and credibility, is of critical importance (Anderson, 2006). Simply gaining media exposure does not guarantee that a claims-maker gains favourable treatment. As Ryan (1991: 53) argues, 'the real battle is over whose interpretation, whose framing of reality, gets the floor'.

Credibility and legitimacy

The news convention of 'objectivity' leads journalists to rely heavily upon news sources. Sigal once observed: 'news is not a reality, but a sampling of sources' portrayals of reality, mediated by news organisations' (1986: 27–8). News articles weave together an assemblage of information attributed to sources that builds up a particular image of the world, leading some sources to gain greater legitimacy than others. There are all sorts of subtle ways in which the credibility of a source can be influenced, including the positioning of their views in the news item, the amount of space they are afforded and any selected accompanying images. When interviewed journalists covering environmental issues present a view of their role as essentially detached (see Chapters 4 and 6). The facts tend to be seen as speaking for themselves and it is up to the reader to interpret them and draw their own conclusions.

Great emphasis is placed upon the balance norm and avoiding advocacy. There are some environmental journalists though who challenge this view and argue that their role should be to assess opposing viewpoints and highlight sources of bias (see Cox, 2013). Also, blogging has opened up a new means whereby journalists can be more open about their personal views.

Importantly the internet potentially allows marginal groups to gain greater control of the framing process and bypass traditional gatekeepers. The shift towards user-generated content and citizen journalism has important implications for the study of sources and the media, and 'alternative journalistic forms radically realign the universe of preferable sources' (Carlson & Franklin, 2011: 5). Also, the use of hyperlinks enables audiences to directly access source material although there is greater scope for mainstream media to supply more contextual material regarding alternative news sources (Atton, 2013). The use of new digital media has offered environmental campaign groups an array of new possibilities in getting their message across. However, for the reasons outlined above, the pressures of the 24/7 news environment with its constant demands for news stories and updates, and little time to verify information, means that journalists frequently fall back on a common stock of predictable, familiar news sources. As Carlson and Franklin point out: 'while new technologies allow non-journalists greater ability to circulate messages, journalists find themselves even more entrenched with tried-and-true sourcing practices in order to meet the unceasing demand for content' (2011: 8).

Controlling framing of movement messages

Environmental NGOs are forced to continually find news angles to get their messages across in mainstream media that often leads them to symbolise their demands through carnival and drama. Gaining and sustaining mass media attention often comes at a price. It is important to recognise that while the web brings with it a host of new possibilities environmental groups are not always seeking to bypass 'old media'. As I argued in Chapter 2 new media tend to be used as an extension rather than replacement of offline media. Studies suggest that once environmentalist voices are accessed there often follows a struggle for control of the news agenda between journalists and environmental campaigners. This is illustrated by Libby Lester's insightful analysis of the Franklin Dam blockade in Tasmania in the 1980s, which resulted in 1,272 people being arrested and 447 put behind bars (Hutchins & Lester, 2006; Lester, 2007). The protest against the proposed construction of a dam in an area of pristine wilderness drew considerable interest from the press in the early stages of the campaign, but it was not long before it was perceived as posing a serious threat to entrenched interests and journalists worked to win back their power. Initially the protest movement had logistical control of the news event but over time elites became more adept at appropriating the symbols that had originally been used by the environmentalists. The protests then became increasingly accused of being 'stage-managed' and attention seeking rather than concerned with the 'real' issues at stake. Hutchins and Lester conclude that since environmentalists' agendas fundamentally challenge the dominant logic of network society, journalists are more likely to treat them differently than political or business news sources and scrutinise their demands more critically (Hutchins & Lester, 2006). This example suggests that environmentalists often experience an uphill battle to get issues framed on their terms and while they may win small victories it is harder to achieve favourable coverage in the long term. Often they may be forced to engage in tactics that result in them losing control of framing the events (Gitlin, 1980; Meyer & Gamson, 1995). Framing is much more difficult for groups that are working against the grain of deep-seated cultural attitudes.

News involves visualising, symbolising and authorising claims, and sources use different discourses of legitimacy based on, for example, scientific or moral appeals (Ericson et al., 1991). Environmental conflicts are frequently portrayed as a morality play of 'goodies' versus 'baddies', David versus Goliath, but environmental claims-makers are often challenged in terms of the scientific legitimacy of their claims. Gamson and Modigliani (1989) draw attention to how issues are organised in interpretive 'media packages' to appeal to news values about what constitutes a newsworthy item that will resonate with publics: 'a package offers a number of different condensing symbols that suggest the core frame and positions in shorthand, making it possible to display the package as a whole with a deft metaphor, catchphrase, or other symbolic devices' (1989: 3). They outline five major devices for framing an issue:

(1) Catchphrases
(2) Metaphors
(3) Exemplars (historical examples from which lessons can be drawn)
(4) Visual images (such as icons)
(5) Depictions (such as moral appeals)

The above 'framing devices' can be usefully examined in the analysis of news media coverage of environmental issues to develop an in-depth appreciation of how particular appeals are constructed. Environmental coverage is not simply about reporting 'the facts'; it involves a narrative interpretation of issues and events that may be underpinned by particular ideological understandings of ecology and society. Some environmental issues resonate powerfully with particular cultural values and beliefs, and may evoke especially strong emotional responses. For example, the Exxon Valdez oil disaster in 1989 where Prince William Sound symbolised for many Alaskans much that they hold dear about their 'unspoiled' natural heritage (see Chapter 5). Framing often has a strong visual component. As Simon Cottle observes:

> Whether depicted in terms of the *spectacularization* of nature as pristine, timeless and outside of human history and society, or through symbolic images of nature as humanly despoiled, exploited and *under threat*, such culturally resonant media images provide an affective charge to environmental discourses and politics of risk now circulating throughout societies.
>
> (2013: 24, original emphasis)

Not all packages resonate equally strongly with wider cultural themes. Nuclear energy and biotechnology draw upon a particularly rich stock of imagery that has strong *cultural resonance* and evokes powerful fears of the unknown (Hansen, 2010). Examples of cultural themes that are often found in biotechnology reporting include, 'Playing God', 'Pandora's Box' and 'Runaway Science' (see Chapter 6). Discourses around biotechnology and nuclear energy tend to draw upon deep-seated fears about the consequences of interfering with nature and anxieties that this may spiral out of control. However, at the same time the 'progress' frame and the 'economic prospects frame' (faith in scientific progress to bring about technological and economic growth) are frequently dominant, especially in the early stages of debate when scientists are often the principal news sources. As Hansen observes:

> Not surprisingly, as it is in the nature of cultural packages that they articulate general ideas and principles rather than issue-specific characteristics, many of the same cultural packages identified in the nuclear power discourse, are equally prevalent in the media and public discourse on genetic engineering and manipulation.
>
> (2010: 112)

These themes of scientific progress and economic prospects can also be identified in recent news media coverage of synthetic biology and nanotechnologies (see Chapter 6). However, over time news frames often shift and new competing ones emerge if a wider range of social actors gain ascendance and an issue is dislodged away from the administrative policy arena to a more overtly political arena (see Chapter 6).

News media coverage is often composed of competing frames and different frames may co-exist together in different public arenas, for example the policy-making arena, or public opinion or media outlets (Gamson & Modigliani, 1989). For example, according to Daley and O'Neill competing frames of the Exxon Valdez oil disaster included: a disaster narrative, a crime narrative and an environmental narrative (see Anderson, 2002). Also, news frames vary across different countries and between different media outlets, genres and conventions (see Boyce & Lewis, 2009; Hansen, 2010). As we shall see in Chapter 5, there are significant differences between the ways in which regional news media may frame environmental issues and national media outlets. A range of studies suggests that non-official news sources may be more likely to gain a greater voice in local news media. For example, Cottle found that local TV news in the UK gave more space to the 'lay' views of ordinary people – Beck's 'voices of the side effects' (Cottle, 2000). Who gets in or on has consequences for how stories are packaged. The voices of ordinary people tend to give greater emphasis to a subjectivity and non-scientific/technical framing.

NGOs often find that they are relatively invisible in mainstream media but social media can provide a valuable means of news entry. For example, Kevin DeLuca and colleagues undertook an analysis of print and online coverage of the Occupy Wall Street demonstrations about rising economic inequality in the US (DeLuca et al., 2012). They found that for the first eight days of the protests there was a complete news blackout in the national daily US press, despite the fact that the events attracted international news coverage. For the first ten days there was also no mention of the protests on any major US television station. It was almost as if the demonstrations did not happen. By contrast from the very outset they received extensive coverage via the internet. A search using Google Blog Search for blogs that mentioned 'Occupy Wall Street' between 17 September and 17 October 2011 came up with more than ten million results (see DeLuca et al., 2012). It wasn't until violence erupted and 700 people were arrested that it finally attracted extensive television news coverage. The twin frames of conflict and negativity permeated mainstream news treatment of the protests. DeLuca argues that two strategies were at work here that sought to marginalise the protestors. The first was to ignore the protests and the second was to frame them negatively.

While the above example does not relate specifically to environmental issues it illustrates a general tendency in traditional media reporting of demonstrations (see Gitlin, 1980). Environmental activists often view mainstream media with fundamental ambivalence. Environmental NGOs are increasingly media savvy and attuned to journalistic news values but playing to these can be risky. At the same time some activists may not fully appreciate how discord and direct action techniques may undermine the message that they are seeking to get across. In the next section we turn to look a little more closely at the often taken for granted judgments that are made when selecting news stories.

News values

Environmental issues do not automatically command news attention. In some instances the news media suddenly pick up on a disaster that has been unfolding for some time (for example, droughts in Africa). In other cases disasters that have received considerable news attention disappear off the news agenda almost as quickly, despite being unresolved, because a new competing item draws greater interest (see Hansen, 2010). Geoffrey Lean, the veteran UK environment correspondent, identifies the following constraints that impact particularly on environmental coverage (see Lean, 1995):

Event-driven reporting

Environmental coverage tends to be highly event-driven. Routine events such as a press conference or international summit provide a peg on which a news story can be hung. Piggybacking on such instances can provide claims-makers with a valuable opportunity to ride on the news attention that they command (Ungar, 1992). More dramatic, what we might call 'trigger events', such as a major hurricane or typhoon, demand global media attention. However, long-running environmental issues that take years to unfold fit uneasily in this formula, and the emphasis upon events rather than underlying causes can result in a narrative that is weak at providing context. Geoffrey Lean observes:

> We are very good at covering *events*, for example, but rather poor at reporting *processes*. So we can handle a Chernobyl or an Exxon Valdez quite well, even though we may not have been very good at exploring the underlying issues that led to both accidents, certainly not in

advance. But we are very poor at reporting such ongoing, imp
processes as desertification, deforestation, or even climate chai

This is often referred to as 'episodic' framing, where coverage is linked
to specific events or particular cases, as opposed to 'thematic' framing
that places issues within a wider socio-political context. Despite such
constraints some intrinsically less dramatic and event-oriented environmental issues have managed to break into the news through clever
packaging on the part of claims-makers.

Image orientation

A second major constraint that in my view has become even more
significant over time is the importance of compelling visual imagery
to convey a story. Environmental news stories that lack strong visual
appeal are considered less newsworthy, especially for television and
online media. Geoffrey Lean observes: 'We are much better at presenting *images* than *arguments*, even in the written media. It often takes an
image to bring an environmental issue alive.' NGOs such as Sea Shepherd Conservation Society and Greenpeace have increasingly geared
themselves towards this new visual culture by providing their own high
quality, dramatic footage (see DeLuca, 1999). Claims-makers also frequently draw upon images that can evoke an emotional response. Over
time powerful associations can be constructed. In the late 1980s dying
seals came to represent pathetic innocence and symbolise an environment increasingly perceived as under threat (see Anderson, 1997). More
recently, polar bears have come to be associated with the effects of global
warming and symbolise threatened pristine environments such as the
Arctic.

Conflict-laden news

A third factor is the news media preference for conflict rather than
consensus. Thus environmental coverage tends to focus on problems
rather than solutions. Heated debates or street battles make for dramatic
coverage. Geoffrey Lean notes:

> We are also much better at reporting *conflict* than the process of
> reaching *consensus*... This is one reason why the media made such a
> mess of reporting the debate over global warming, giving equal space
> and weight to a handful of dissident scientists as to the overwhelming
> consensus of the vast majority.

Objectivity and balance

Journalists are frequently attacked for their reporting of controversial environmental issues and scientists often accuse them of sensationalism and inaccurate reporting of the 'facts'. It is often simply assumed that this is the result of a lack of scientific training or a deliberate attempt to distort information. It is rarely as simple as this. As I have argued, all competing accounts of reality are inevitably partial as they highlight certain aspects over others. Scientists simplify their work in presenting it to a wide audience by frequently using metaphors and other devices that are designed to garner acceptance of their work. Equally other news sources such as politicians or environmental groups attempt to spin their material in particular ways. Environmental issues are often hotly contested and as Anders Hansen remarks: 'journalists have a difficult balancing act to perform amid a barrage of conflicting claims about environmental issues' (2010: 90).

Professional ideologies around 'objectivity' and 'impartiality' permeate journalism. When interviewed environmental correspondents often stress their detached role in reporting a subject area now increasingly politicised and fraught with controversy. For example, when asked 'how do you regard the environmental pressure groups as sources?' ITN's Science Editor told me:

> Increasingly tainted. Partly because they've become more shrill and more extreme as the mainstream political parties have stolen their agendas. That's pushed them into a corner, often where they're having to argue for extreme solutions and I think that means that people who are editing programmes are more suspicious of them and less trusting. And they have also become a business, their turnover is bigger than ITN's.
>
> (Interview, 1993)

Adherence to the *balance norm*, the idea of giving different sides of the debate an equal hearing has been influential in the past, but can be attacked on the grounds of giving the impression that a scientific consensus has yet to be achieved (see Boykoff & Boykoff, 2004). A former Environment Correspondent for ITN claimed:

> It's not my business to ram the environment down the public's throat in a prejudicial way ... The role of the media, any decent journalist is to tell them this is the other side of things ... so you never run a biased prejudicial strong environmental message because ... anybody in the world can make up and invent or get some facts to support

a strong environmental message ... So yes you can use a strong environmental message as an initial foot if it is well sourced and well researched, but you never put one out, just one side of a message.

(Interview, 1993)

Nowhere is this more controversial perhaps than with the heated topic of climate change. There have been several cases where journalistic impartiality has been called into question in relation to climate change and, recalling Nick Davies' ten rules of news selection discussed earlier, it has become an increasingly 'risky' story for reporters to cover. The threat of what Herman and Chomsky refer to as 'flak' – negative responses to media coverage that may take the form of letters, phone calls, social media campaigns, petitions or lawsuits – can be a powerful disincentive to unsettling the status quo (Herman & Chomsky, 2002).

For example, in September 2007 the UK public service broadcaster, the BBC, decided to scrap a planned climate change special TV show, *Planet Relief*, as it was seen as too campaigning and potentially leaving the corporation open to criticisms of bias (Black, 2007). Clearly the BBC perceived this form of Comic Relief type event as too political, given that annual public awareness events on global justice/poverty have not attracted the same kind of reaction. The decision to drop the day-long special came after the comments of Head of BBC News, Peter Horrocks, at the International Edinburgh Festival, when he was questioned as to whether the BBC should campaign on issues such as climate change. He remarked in no uncertain terms: 'I absolutely don't think we should do that, because it's not impartial. It's not our job to lead people and proselytise about it ... It is absolutely not the BBC's job to save the planet. I think there are a lot of people who think that, but it must be stopped' (Plunkett, 2007). Following mounting criticism, much of it from organised deniers, in 2010 it was announced that the BBC Trust was to launch an investigation into allegations of bias in its coverage of science including the reporting of climate change (see Jones, 2011). Following the publication of the report some UK newspapers, notably the right-leaning *Daily Mail*, claimed that the report itself was biased – against climate change sceptics.

Why we should give the cold shoulder to a BBC trust review that argues the broadcaster should ignore global-warming 'deniers'. *Daily Mail* 24 July 2011

Later that year the Global Warming Policy Foundation, an influential body of climate change sceptics, published the report 'The BBC and Climate Change: A Triple Betrayal' claiming:

> the BBC has not only failed in its professional duty to report fully and accurately on one of the biggest scientific and political stories of our time: it has betrayed its own principles... On one of the most important and far-reaching issues of our time, its coverage has been so tendentious that it has given its viewers a picture not just misleading but at times even fraudulent.
>
> (Booker, 2011: 9–10)

In December 2012 the Global Warming Policy Foundation sent an open letter to the BBC Director General calling for a new high-level seminar to address the issues. The BBC implemented many of the recommendations from the BBC Trust report, including greater training for journalists on impartiality, but did not act on the concern that there was an insistence on bringing in dissident voices despite the 'non-contentious' nature of some stories (see Jones, 2011; Harvey, 2013). At the time of writing the corporation is facing new accusations of giving undue weight to climate change sceptics (Harvey, 2013; Vidal, 2014).

A climate of fear also seems to have impacted elsewhere. For example, in 2009 the news agency Reuters launched a Facebook page calling for donations to 'Help Reuters Stop Global Warming'. However, by 2011 a climate change sceptic, Paul Ingrassia, had joined the organisation as a senior editor and the US Media Matters for America Advocacy group claimed that its survey of reporting over the six months immediately prior to Ingrassia's appointment compared with an analogous period in 2012 revealed a 48 per cent drop in levels of coverage. Veteran reporter David Fogarty, who had been covering climate change at Reuters for four and a half years, claimed that by early 2012:

> Progressively, getting any climate change-themed story published got harder. It was a lottery. Some desk editors happily subbed and pushed the button. Others agonised and asked a million questions. Debate on some story ideas generated endless bureaucracy by editors frightened to take a decision, reflecting a different type of climate within Reuters – the climate of fear.
>
> (Goldenberg, 2013)

Advertising can also act as a major constraint upon environmental coverage. Some news outlets devote a significant amount of space to adverts for oil companies, air travel and cars (see Monbiot, 2007). Dependency on advertising revenue from fossil fuel industries may lead journalists to self-censor stories so that they do not directly threaten their interests. Gelbspan (2005) documents the case of a television editor who, on carrying a report linking a huge flood with climate change, was threatened that oil and car advertising would be withdrawn.

Covering controversial environmental issues in the mainstream media then often puts journalists in a difficult position. Topics such as climate change are complex and contested, and challenge deep vested interests. In factual news reporting journalists emphasise the need to be neutral and open-minded, and it is rare to find any acknowledgement that environmental coverage is inevitably shaped to some extent by subjective factors (Anderson, 1997; Chapman et al., 1997). While mainstream media are likely to avoid reporting which fundamentally challenges corporate interests that have a strong connection with their outlets, it would be too simplistic to suggest that this completely prevents critical reporting. As Carvalho (2007) observes, there is always some space for a diversity of opinions to be aired. Indeed, investigative coverage by such journalists as *New York Times* Environment Editor Andy Revkin and UK *Guardian* columnist George Monbiot has played a major role in exposing the politics of climate-science scepticism. For example, in 2005 Andy Revkin uncovered links between the climate denial industry and the White House. A *New York Times* investigation found that Phillip Cooney, an ex lobbyist for the American Petroleum Industry and a lawyer with no scientific training, frequently edited government climate reports to highlight doubts about the connection between climate change and fossil fuel emissions (Monbiot, 2006; Revkin, 2005).

However, in some regions of the world journalists covering environmental affairs may face great personal threats that put their lives at risk. Reporters without Borders, a non-profit organisation that campaigns for greater journalistic freedoms claims: 'More and more journalists are nowadays being killed, imprisoned, harassed, censored or intimidated because of their coverage of environmental issues' (Reporters without Borders, 2009). In September 2012 environmental journalist Hang Serie Oudom was found murdered in Cambodia less than a week after the publication of the latest in a series of reports exposing illegal logging. Another example concerns Lucio Flavio Pinto, award winning journalist and founder of the independent magazine *Jornal Pessoal* in Brazil, who has been the victim of much intimidation. A number of lawsuits were launched against him after his coverage of deforestation in the Amazon.

Summary

Far from mirroring reality the coverage of environmental issues, as with news in general, is highly selective and reflects economic, political and cultural factors. News about the environment is the end product of a complex process of construction. The study of the relationship between news sources and the media remains as important as ever. The digital revolution has brought about new layers of complexity but has not completely transformed earlier patterns of dominance. The preceding discussion suggests that news media ownership and control, and patterns of source dependence can exert an important influence on the framing of environmental issues. Editors' decisions may be influenced by the fear that running critical items may result in lost advertising revenue or result in 'flak'. Also, economic conditions can impact on the capacity of journalists to undertake in-depth investigative reporting. There is little evidence that we have entered into an era of minimal news media agenda-setting effects, although the digital revolution does seem to have led to more narrowcasting. However, it would be far too simplistic to suggest that media content can simply be viewed as determined by economic factors alone.

What gets in the news and whose voices gain most prominence or legitimacy is the outcome of a complex series of internal and external factors. As we shall see in the chapters that follow this varies depending on the issue and can shift over time.

Further reading

Carlson (2009) 'Dueling, Dancing, or Dominating? Journalists and their Sources', *Sociology Compass*, 3 (4), 526–542.
Cox, R. (2013) *Environmental Communication and the Public Sphere*. London: Sage. Chapter 6.
Entman, R. M. (1993) 'Framing: Towards Clarification of a Fractured Paradigm', *Journal of Communication*, 43 (4), 51–58.
Hansen, A. (2010) *Environment, Media and Communication*. London: Routledge. Chapter 2.

4
The Climate Change Controversy

> I think climate change is the biggest under-reported, or unre-
> ported story of our times...and yet if you look around, then
> the critical evidence suggests that not only are the effects of cli-
> mate change already being felt, they are (worsening) within the
> next few years.
>
> (Daniel Kalinaki, cited in Corner, 2011: 19)

Climate change is the most serious issue of our time and yet it is
deeply contested. Journalists face significant challenges in sustaining
coverage of this complex field of science. Once at pains to state their
role as 'impartial' commentators, media professionals have become
increasingly embroiled in the debate. This chapter examines how climate change has been framed over time within the news media and
which voices have been treated as credible and authoritative sources.
The media, particularly television, provide a key source of information
for publics on climate science (Boykoff & Boykoff, 2007). Important
questions are raised concerning objectivity and trust in the communication of controversial science. The news media have been blamed
both for exaggerating *and* underestimating the risks of climate change.
This chapter explores how far the news media have faithfully reported
climate science raising important questions concerning objectivity and
trust in the communication of controversial science. It surveys research
conducted in a range of international contexts, supplemented by the
author's own findings drawn from an analysis of UK national press
coverage of the Rio+20 summit in 2012, and interviews with national
print journalists and broadcasters. The analysis suggests that there is
little direct correspondence between scientific knowledge of climate
change, and prominence and framing of climate science within the

news media. Social, political and economic factors are crucial when seeking to understand patterns of coverage over time. Climate science has become increasingly politicised and the news media, far from detached bystanders, have played a key role in influencing the contours of the debate. There are a number of key questions this chapter sets out to explore. What factors account for the peaks and troughs of levels of news media attention and public concern over time? Is there a close correspondence between the two and how far are they related? What role has the economic recession played in the recent trend in Western countries towards generally declining levels of concern over climate change and increasing scepticism? To what degree have national political agendas affected patterns of media attention? How much power do editors have to shape the agenda? To what extent have issue-entrepreneurs (including climate change think tanks and non-governmental sources) shaped the debate? What role have extreme weather events played? And how far is this explained by traditional issue-attention cycle models?

As will become apparent, none of these questions are straightforward to answer. Most research to date has focussed for practical reasons on the press, particularly elite newspapers. Also, longitudinal studies of press reporting which seek to track levels of attention over time across a number of different countries tend to take single newspapers as a proxy for press coverage in a single country. However, previous studies suggest that there are significant differences in the amount of coverage between different newspapers in the same country, as well as important variations in ideological positioning (Carvalho, 2007).

A new culture of nature?

As we heard in Chapter 2, Castells views the environmental movement as 'one of the most decisive social movements of our time' (2009: 321). In *Communication Power*, which was first published in 2009, he argues that public concern about climate change has rapidly grown 'at an unprecedented level' across the globe (2009: 317). He sees the media as having played a key role in bringing about such shifts in attitudes observing that: 'people make up their minds according to the images and information they retrieve from communication networks, among which the mass media were the primary source for the majority of citizens during the two decades when awareness of global warming increased' (2009: 315). The internet contains a deluge of information on climate change; Schäfer found 377 million hits alone returned for 'climate change' in an English language Google search, even surpassing the number of hits for 'Barack Obama' (Schäfer, 2012). What brought this greening of the media about? For Castells it was the networking between environmentalists, scientists and celebrities that catapulted the issues into the media spotlight. Moreover, new communication technologies, he observes, have helped environmental organisations mobilise supporters through their websites, YouTube videos, email and social networking sites such as Facebook. Changes in the public mind are seen as having brought about policy changes and 'The social movement to control climate change has largely succeeded in raising awareness and inducing policy measures,' although he adds the caveat 'albeit woefully inadequate to this point' (2009: 337). For Castells, then, this alliance of social actors aided by the networking power of the internet has led to a transformation in the way that we view nature. But does this gloss over some of the complexities of the processes of power and counter power? In order to answer this question we need to look a little more closely at the peaks and troughs of media and public interest in climate change over time.

Over the past decade climate change has gained greater prominence on local and national policy agendas across the world and yet government leaders have been slow to act and public attitudes about the reality, scale and urgency of the issues have dipped and wavered (Moser & Dilling, 2011). As we heard in Chapter 1, while the links to the frequency and intensity of extreme weather events, the rate of change and the seriousness of the threat is the subject of much debate, there is a clear scientific consensus that the mean global temperature has increased significantly and human activity is the main cause (see Brulle et al., 2012). Yet in the US and, to a lesser extent, in the UK organised groups of right-wing think tanks and climate sceptics have used a range of communication outlets to steadily trickle-feed seeds of doubt and attack individual scientists. It is evident that over the past decade significant ideological and partisan polarisation has occurred on the issue among the US public (see McCright & Dunlap, 2012; Rolfe-Redding, et al., 2011). Those who believe that scientists disagree over climate change tend to be less convinced that it is taking place and less likely to support climate policy (see Ding et al., 2011).

I start by briefly sketching the early history of climate change reporting before moving on to consider more recent developments in greater depth. The focus here is particularly on the UK and the US, since the bulk of research to date has concentrated on these two countries. As we

shall see, it has taken a relatively lengthy time for the media to communicate the scientific consensus that has now been reached that climate
change is real.

The early history of climate change reporting

Late 1980s: Green issues rise up the political agenda

Up until the mid 1980s there was generally little discussion of climate change/global warming within the national print media, although
Boykoff and Rajan (2007) identify a few isolated articles in the US and
UK national press exploring human contributions as far back as the
1950s. The first sharp spike in coverage in the British and US national
daily newspapers first occurred in the late 1980s as the issues became
increasingly politicised (see Boykoff & Boykoff, 2004; Boykoff & Rajan,
2007; Carvalho & Burgess, 2005; Wilkins, 1993; Zehr, 2000). During a
drought and record high temperatures in the summer of 1988 James
Hansen testified to Congress that he was 99 per cent confident that a
long-term trend towards global warming was occurring. The US opinion-
leading newspapers leapt on the story and in the space of a year the
number of news articles about global warming increased tenfold (Ingram
et al., 1992; Weart, 2008). At the same time public opinion polls indi-
cated that there was rising concern about the 'greenhouse effect' (see
Weart, 2008). That September Margaret Thatcher, the former British
Prime Minister, made a 'green' speech to the Royal Society propelling
the issue further up the agenda (see Anderson, 1997).

  During the mid to late 1980s a scientific discourse dominated with climate scientists appearing as the main news sources in the UK national
press. While newspaper articles generally took the view that global
warming existed, the risks tended to be downplayed and it was framed
within narrow scientific terms, with few articles delving into the wider
socio-political dimensions (see Carvalho & Burgess, 2005). However, following Margaret Thatcher's intervention scientists were no longer the
exclusive definers of the debate as political actors increasingly attempted
to shape the agenda (Carvalho & Burgess, 2005). A similar shift over this
period also took place in the agenda-setting US and German national
newspapers (Trumbo, 1996; Weingart et al., 2000). The topic had moved
from being principally covered as a science feature story to a news item
centring on political controversy (Wilkins & Patterson, 1991). Responding to the rising salience of the issues UK television news began to
appoint its own specialist environmental correspondents. One former
ITN Environment Correspondent observed:

It was very fashionable in the late 1980s and because a lot of these issues hadn't really been covered...and a lot of them were also new stories. There was a whole question mark over whether global warming existed or the hole in the ozone existed and all these sort of things...so there was all this thing in the late 1980s and early 90s, a lot of interest, mainly among the news editors. So once it was proved there was global warming and once the initial breaking of new ground, and there were a hell of a lot of issues which had been building up behind a dam for what it felt a long time, suddenly came out and environmental groups were getting very big and wealthy and therefore had more resources to target the media and target the news.

(Interview, 1993)

However, this flurry of news interest in the 'greenhouse effect' and environmental issues more generally was soon to be overtaken by what were perceived to be the more immediate issues of the recession and the Gulf War.

Early to mid-1990s: Politicians waver and editorial fatigue sets in

Following the significant rise in interest in the 1980s there was a marked decline in national press coverage in the UK and US between 1991 and 1996 as political interest waned, despite a brief blip in 1995 at the time of the release of the IPCC report (Anderson & Gaber, 1993a, 1993b; McComas & Shanahan, 1999; Ungar, 1992). Also, at this time the US media started to pay more attention to the various conservative think tanks that started to mobilise around climate change (McCright & Dunlap, 2003). Several factors contributed to this steep fall including political inertia, the problems journalists encountered in sustaining the newsworthiness of the issue, counting the costs of taking action and the impact of the economic recession. Interviewed in 1993, a former science correspondent for BBC News stated: 'I think basically the stuffing has gone out of the story. It's no longer the major political issue it once was.' A former Environment Correspondent for Channel Four News concurred, noting:

While Thatcher was perceived to be keen on the environment and while there was a Conservative thought going around that we own the earth therefore we've got to do something about it...you got a lot of coverage in the Conservative papers. When that coverage came to, as it was bound to be, seen to be dangerous to financial interests, then the enthusiasm declined suddenly. In other words, as soon as the environment correspondents dug in and started looking for the real causes of things ... editors were actually slightly less keen on running their stories.

(Interview, 1993)

Similarly, a former ITN environment correspondent observed:

Then John Major replaced Thatcher and he was perceived as being less interested in green issues and the news desks started saying 'oh no, not another story on global warming'. They saw it as a one-off story or a two-off story ... in the beginning, you know, the whole question was does global warming exist or not. Once you'd really had a 'yes' then really that's it, we're all doomed. And so they never really were interested in going further than that. And also pictorially it's very difficult. Not so much in print I guess but pictorially it's a tricky one to show because obviously they're showing something of the future. You cannot show what it is now apart from a few graphics. Once you've done those fancy graphics and you've shown a few bits of East Anglia, you know, being battered by rising sea levels ... I mean there is a sort of limit before you've really got to start to use the same pictures again, but different script, different words.

(Interview, 1993)

The long-term nature of global warming and lack of visibility clearly contributed to problems in sustaining media interest over time. A former Environment Correspondent for BBC News observed: 'Above all environmental stories really need good pictures ... global warming is very difficult because you can't actually see global warming' (Interview, 1990, see Anderson, 1997: 121–2).

One of the most widely used images to illustrate climate change has been melting polar ice caps (Doyle, 2011). Central to the problem of representing climate change is the role of the visual (photographic and pictorial evidence), the spatial (geographical distance) and the temporal (the time lag). As Doyle argues:

polar ice caps function as privileged signs within the discourse of climate change ... Whilst undoubtedly providing an important impetus to the politics of climate change, at the same time such images produce a distancing effect, relegating climate change impacts to a remote and inaccessible place where animals and habitats are affected rather than humans.

(Doyle, 2007: 142)

However, while media interest in climate change generally went off the boil, there was a significant rise in the number of articles being published on this topic in scientific journals such as *New Scientist* and *Nature* (Weart, 2008).

Mid-1990s to early 2000s: Floods, flowering grasses and freezing rain – climate science breaks through

By the mid-1990s further scientific evidence of global warming was accumulating as British Antarctic Survey researchers discovered that a flowering grass was now 25 times more common compared with 30 years ago and the IPCC proclaimed that the world's climate was still 'at serious risk'. Geoffrey Lean, the then environment correspondent at the *Independent on Sunday*, viewed this as a breakthrough in the previous news cycle:

> If they're not in the public arena it's always very hard these things to judge because the press ownership is rather insecure and so it tends by in large to write what the press... it tends to report what everybody else is reporting, what the politicians are interested in, and the politicians are interested in what the press is interested in and there are a lot of self-reinforcing circles and it's quite hard to break into those circles. Now an event can break into it, a major event can break into it... on climate change, for example... some major discovery about what's going on with the climate can break into it and indeed it did. There was a great quietness about global warming issues in the press for a long time, and then the beginning of flowering on the Antarctic Peninsula sort of broke the cycle.
>
> (Interview, 2006)

A second spike in newspaper coverage occurred between 1997 and 2003, as the effects of climate change were brought closer to home (Boykoff & Rajan, 2007; Boykoff & Boykoff, 2007; Carvalho & Burgess, 2005). Important policy events such as the Kyoto Climate Summit in 1997, together with extreme weather events (including persistent freezing rain in the US and Canada in 1998, and the floods in Asia in 2002) clearly generated more coverage (see Boykoff & Boykoff, 2007; Carvalho & Burgess, 2005). Over 3,500 journalists and in excess of

400 media organisations from 160 countries were involved in covering Kyoto (Leggett, 2001 cited in Boykoff & Roberts, 2007). During this period there is much less research evidence about television news coverage. However, one study based upon data from the Vanderbilt Television News Archive in the US found a relatively small number of stories compared with the 1970s and the 1980s; the three US television networks broadcast approximately 100 news stories on climate change over the period 1990 to 1999 (Sachsman, 2000).

Mid-2000s to late 2000s: Near scientific consensus reached

The largest increase in news media attention to climate change to date occurred between 2003 and 2007 in part generated by Hurricane Katrina, the release of Al Gore's film, *An Inconvenient Truth*, and the Fourth IPCC Assessment Report. Climate change rose on the political agenda with the G8 Summit in 2005 and the coming into force of the Kyoto Protocol. This led news editors, who tend to come up through a political journalism route, to give the topic more priority. The March 2006 edition of *Time* magazine reported:

> Never mind what you've heard about global warming as a slow-motion emergency that would take decades to play out. Suddenly and unexpectedly, the crisis is upon us. From heat waves to storms to floods to fires to massive glacial melts, the global climate seems to be crashing around us.

Almost a year later, in April 2007, it launched its 'Global Warming Survival Guide'. Over this period there was a steady rise in attention among US agenda-setting newspapers and an increase in coverage in Australia/New Zealand, the Middle East, Asia, Eastern Europe and South Africa, but a more significant increase in the UK (Boykoff, 2007a; Boykoff & Roberts, 2007: 39–41; Speck, 2010). Ironically there tends to be comparatively little media reporting on climate change in developing countries, yet they are likely to suffer the worst effects (Painter, 2007). In 2006 there was four times as much coverage of climate change issues in the UK elite national newspapers than there was in 2003. While in the US over the same time span coverage increased in the prestige press by approximately two and a half times (Boykoff, 2007b). In the UK many of the popular national newspapers also started to give the issues greater prominence. For example, in September 2006 the UK's highest circulating daily newspaper, *The Sun*, launched its own campaign on climate change urging readers to 'Go Green with the Sun'.

Previously the newspaper's coverage of the issues was sporadic and low key (Gavin, 2007).

Environmental reporting in the UK had clearly moved from being a specialist scientific field to an all encompassing issue that could fit under virtually any area of news coverage. John Vidal of the UK's *Guardian* newspaper observed:

> Environment journalism has come a long way since 1975 when Geoffrey Lean – then of the Observer, now of the Telegraph – became the first dedicated correspondent. Before that, the brief was mostly given to correspondents who shadowed the government's rural affairs or farming department. The beat still covers traditional areas such as floods, spuds and trees, but it is now centred on science writing, international development and politics, energy, technology, economics, celebrity and lifestyle, as well as business, trade and protest. And because it crosses so many traditional journalistic boundaries, it has become a specialist area that suits generalists. Equally, there is no specialist political, business or feature writer who does not now regularly report on the environment. To paraphrase Al Gore, we are all environment journalists now.
>
> (Vidal, 2009)

In the UK context the passing of the Climate Change Act in 2008 meant that the topic was clearly part of a permanent agenda rather than a fleeting fashion (Philo & Happer, 2013). Across many western countries the issues had moved from specialist environment sections to a wide spread of news beats. A scientific frame no longer dominated and the subject matter was just as likely to be covered using a business/policy frame (Anderson, 2009; Sampei & Aoyagi-Usui, 2009; Young & Dugas, 2011).

To briefly summarise then, over the course of these four time periods a number of important changes to climate science reporting, and environmental journalism more broadly, occurred. What we see are dynamic, shifting processes influenced by multiple factors including the complexity of the issues, competing news stories, issue saturation and novelty, event-centred reporting, ideological conflict and resource mobilisation/issue subsidies as well as the rise of digital communication technologies.

Firstly, the volume of newspaper reporting substantially increased by 2007 – at least in many parts of Europe and the US. Indeed, a recent study suggests that worldwide press coverage of climate change increased between four and eight fold over the time period 1996 to 2010 (Schäfer et al., 2011). However, it is important to note that there have been a number of peaks and troughs over time; for example in the early 1990s coverage substantially dipped. Secondly, the impact of climate change began to become more visible and near scientific consensus was reached. Thus objective indicators about the evidence of climate change were also important factors. Thirdly, scientists were no longer the exclusive definers of the debate, as political actors became increasingly significant players. At the same time, a much wider range of claims-makers (including NGOs and celebrities) became more prominent voices within the news media. And, finally, environmental issues became increasingly politicised and coverage moved from being viewed as a specialist area of reporting to being seen as impacting upon several different journalistic beats. However, in 2012 despite Hurricane Sandy, a melting ice cap and a worldwide drought climate change failed to surface very prominently in news outlets around the developed world tracked by Daily Climate (Fischer, 2013).

A number of different explanations have been offered in an attempt to understand these shifts. Perhaps the most influential model used to explain the rise and fall of news media interest in environmental issues is Downs' 'issue-attention cycle' to which we turn to examine next. Downs (1972) suggests that social issues tend to go through a cycle of fervent interest followed by growing boredom. To what extent does this apply to climate change?

Downs' issue-attention cycle

The American sociologist Anthony Downs postulated that characteristically social problems could be observed to go through the following five stages in a cycle of increasing media interest followed by a relatively swift decline in attention (see Figure 4.1).

In general Downs argues that an issue is likely to fade from media interest if its dramatic/entertainment value decreases, if it no longer affects everyone or if it is no longer in the interests of power holders in society and will involve major upheaval and costs. However, he predicted that the intrinsic character of environmental issues meant they are unlikely to quickly enter the 'post-problem' stage. That is, they tend to be more visible and threatening than other social issues; they encompass a wide range of causes and override political divides; many environmental issues can be solved through technological means; a small group in society (industry) can be blamed and companies can profit from environmental services and products (see Anderson, 1997).

Stage One: *The pre-problem stage*

The first stage in the cycle is labelled the 'pre-problem' stage; here a problem exists but it has yet to spark attention.

Stage Two: *Alarmed discovery and euphoric enthusiasm*

The second phase is where the issue becomes regarded as a hot topic and dramatically leaps into the media spotlight causing public alarm.

Stage Three: *Realising the cost of significant progress*

The third stage is where coverage becomes more subdued as people recognise that it will involve significant sacrifices (for example, financial resources, changes in behaviour and so on).

Stage Four: *Gradual decline of intense public interest*

Next public interest typically starts to wane and the issue comes to be seen as less newsworthy.

Stage Five: *The post-problem stage*

In the final stage of the cycle, attention towards the issue settles down and there is increasing boredom although the problem itself remains unresolved.

Downs (1972) 'Up and down with ecology – the 'issue-attention cycle'', *Public Interest*, 28: 28–50

Figure 4.1 Downs' issue-attention cycle

The difficulty with Downs' model is that it is overly linear and inflexible. Also, it focuses upon a limited number of fora, or arenas of influence, namely: the media and public agendas (Hilgartner & Bosk, 1988). This 'one model fits all' approach tends to underplay the complexity of public debates. An extremely diverse range of problems may be included under the heading of 'environmental' issues and, within cycles of issue attention, their fate is determined by a complex array of factors (see Anderson, 1997: 30–2). The model fails to account for differing levels of attention to an issue accorded by different media outlets. Rather than climate change overriding political divides it has become deeply divisive. Moreover, studies suggest that the model may apply differently depending upon the cultural context (Brossard et al., 2004). The role of political institutions, NGOs, the wider political culture and the scientific community has been demonstrated to play a key role in defining the important issues of the moment. None of these public arenas are discrete units; they encompass a broad range of overlapping platforms employing diverse strategies and targeting different audiences.

Similarly Castells' explanation for the greening of the media tends to be overly linear and based upon the experience of Western countries. Also, he makes some rather sweeping claims about changes in public attitudes over time. Castells claims: 'from the late 1980s to the late 2000s there has been a dramatic shift in the world's public opinion in terms of awareness and concern regarding its potential consequences' (2009: 315). As we shall see later on in this chapter this overlooks significant cross-cultural differences in opinion about climate science. First I want to look a little more closely at differences in news media attention to climate change across the globe.

Differential patterns of attention to climate change

While Castells' broad-brush analysis provides us with a generally useful account of the rise of climate change within the news media, it tends to be very much centred upon the experience of Western cultures. A more nuanced analysis suggests that there is considerable variation between countries – both in the amount of coverage and in its framing. Castells gives insufficient attention to how different geographical and cultural values shape issue-attention cycles and different news frames.

First we can observe major differences in the *amount* of press coverage of climate change around the world. Max Boykoff and Maria Mansfield have tracked newspaper coverage of climate change since 2004. The graph below (see Figure 4.2) has to be interpreted with some caution since it is not possible to directly compare the volume of coverage between different countries as it mainly includes English language publications and the newspapers are not evenly distributed across the different continents (Painter, 2010: 15). Indeed, the problem with large-scale comparative research into media coverage of climate change, thus far, is that it is difficult to draw many meaningful conclusions about different continents given small sample sizes and the fact that it tends to be based simply on an analysis of newspaper coverage.

The graph suggests that there has been significantly less press coverage of climate change in South America and Africa over time. However, it needs to be borne in mind that despite the digital revolution and the growth of mobile phone use, radio is still the most popular source of information in Africa (see Painter, 2010). Also, can we take two newspapers, *Business Day* and *The Financial Mail* (both South African newspapers), as a proxy for coverage in the whole of the continent of Africa? Similarly, can we take one Argentinian newspaper, *Clarin*, to represent

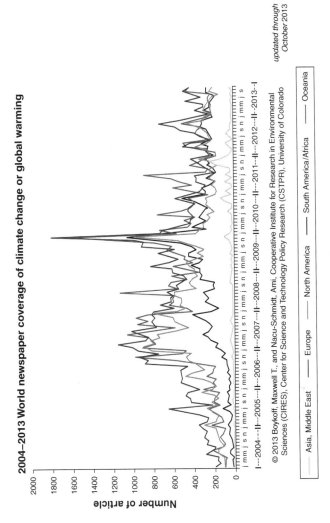

Figure 4.2 World newspaper coverage of climate change or global warming
Source: Reproduced with kind permission by Max Boykoff and Maria Mansfield.

the whole of South America? For practical reasons, including language barriers, most research into media coverage of climate change is disproportionately based upon press coverage in the UK and the US. This is a major limitation given that television is one of the most important sources of information on climate change. As Painter observes:

> There is evidence to suggest that national television is not just the most used but the most *trusted* source of news in most countries by some distance, when compared to newspapers, news websites, radio and blogs. But there are several practical obstacles to including TV broadcasts in any content analysis. It is immensely time-consuming to watch television broadcasts or listen to radio programmes compared to processing print and online media. It is also difficult to physically get hold of recordings of the programmes in sufficient quantities to make the sample robust.
>
> (2010: 13–14)

Research also indicates that there can be wide variations in the levels of attention devoted to climate change between different media outlets reflecting different ideological positions. For example, a study of the UK national press coverage of the Rio+20 Earth Summit in 2012 illustrates how *The Guardian* newspaper published by far the most articles especially in comparison to the popular mass circulation newspapers, most of whom only carried a handful of items – see Figure 4.3 below. Indeed, *The Guardian* was the only major news

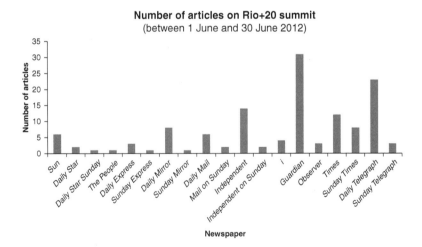

Figure 4.3   Number of articles on Rio+20 summit

organisation to set up a webpage devoted to the summit. The newspaper also published a live daily blog that updated things as they happened and provided Twitter hash-tags and video-links (Fahn, 2012). *The Guardian* is known to devote a comparatively large amount of space to environmental issues compared with other UK national newspapers; it exhibits a social democratic ideology and often voices the values of solidarity, justice and global responsibility (see Carvalho, 2007).

Unfortunately, there are many methodological challenges that currently beset attempts to map global coverage of climate change. For example, Boykoff and Mansfield take 16 newspapers (including Sunday versions) to represent the small island of the UK and six newspapers to represent the US but, as mentioned above, only one newspaper is used to represent the continent of South America and only two newspapers to represent the whole of Africa. According to the World Bank, African communities, particularly the poor in coastal cities and on low-lying islands, are among the world's most vulnerable people to climate change. We need to know much more about how the issues are being communicated in African countries through examining a range of media outlets (World Bank, 2013).

With these considerable limitations in mind Figure 4.2 suggests a significant peak towards the end of 2009 at the time of the "Climategate" scandal and the Copenhagen COP-15 climate talks, followed by a marked decline going back to similar levels to those in 2005. This appears to have been followed by a slight upward trend more recently, with the exception of South America and Africa. However, single country studies of press coverage reveal further complexities. For example, a recent South American study by Bruno Takahashi and Mark Meisner examined climate change coverage in ten Peruvian newspapers between January 2000 and December 2010 (Takahashi & Meisner, 2013). Their content analysis suggests a rather different pattern of coverage with the sharpest increase in volume of articles occurring between 2007 and 2008.

However, despite country differences the general spike in press attention to climate change at the end of 2009 identified by Boykoff and Mansfield is confirmed by other research. For example, Fischer (2011) reports that an analysis of DailyClimate.org's archives of media coverage of climate change across the globe reveals journalists published 23,156 climate related stories in English in 2010 – a 30 per cent drop from the number of stories in 2009. Despite this general fall particular news outlets bucked the trend including Reuters, the *New York Times*, *The Guardian* and the Associated Press. Moreover, research undertaken

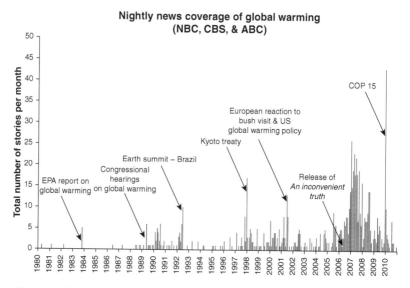

Figure 4.4 Evening news coverage of global warming in the US
Source: Reproduced with kind permission from Robert Brulle.

by Robert Brulle that has tracked nightly television news coverage (NBC, CBS and ABC) of climate change in the US since the 1980s found a similar trend (see Revkin, 2011) – see Figure 4.4.

More recently, a large-scale longitudinal study has been undertaken at the University of Hamburg that maps climate change coverage in leading print media from 23 countries between 1996 and 2010 (Schäfer et al., 2011). Again, as the authors acknowledge, there are issues concerning the extent to which generalisations can be drawn. The authors took one leading newspaper from each of the selected countries around the globe and measured the amount of climate change coverage as a proportion of the entire coverage in that publication per month. The findings suggest that there was little coverage of climate change in the African press over this time period. The problem with this of course is that the chosen newspaper may not be particularly representative of news media coverage as a whole in that country. Indeed, studies suggest that there are sometimes wide variations between newspapers in a single country, both the amount of space devoted to climate change and the framing of the issues (Carvalho, 2007). Nevertheless this finding is supported by the conclusions of other pieces of research. A more detailed study of newspapers in South Africa, Nigeria and Ghana by Evelyn Tagbo (2010)

concluded that coverage was generally very low and h
dent on international news agencies. As Mike Shanahan,
at the International Institute for Environment and Develop
observes:

> The general picture painted by the most recent research is   ...mle
> coverage of climate change in non-industrialised countries is increas-
> ing, the quantity and quality of the reporting do not match the scale
> of the problem. The studies reveal a reliance on reports from Western
> news agencies rather than more locally relevant news ... Editors tend
> to give climate change a low priority.
>
> (2009: 156)

Mike Shanahan identifies six main barriers to the coverage of climate
change in non-industrialised countries (Figure 4.5):

---

1. *Lack of training for journalists to make climate change relevant and interesting:* Climate change is an immense topic, often feels remote, and involves complex jargon and difficult concepts to describe like risk and uncertainty.
2. *Lack of access to specialist knowledge* in some developing countries. Much of the science is Western and written in English. Many local scientists lack awareness of media needs such as clarity and brevity, whilst government officials who are specialists on climate change are often unavailable.
3. *Lack of resources to travel* outside city centres to garner first-hand evidence and testimony or to attend international climate conferences.
4. *Lack of research available on the country-specific current and future impacts* of climate change.
5. *Too cosy a relationship* between the media and national political leaders, which can lead to an unquestioning attitude of the powers-that-be or over-willingness to blame the West for global warming.
6. *Unsympathetic editors* who do not see the importance of the story compared to other more pressing or popular concerns.

Figure 4.5 Major barriers to reporting climate change in non-industrialised countries
Source: Mike Shanahan in Painter, J. (2010) *Summoned by Science*, p.20
Reproduced with kind permission from James Painter.

As well as considerable differences in the quantity of coverage there
are significant variations in how climate change is framed. For example, Dominique Brossard and colleagues (2004) examined newspaper
coverage of climate change from 1987 to 1997 and found that French
coverage was focused on international relations, while US coverage
centred on conflict between scientists and politicians.

Another example is the work of Lisa Antilla who examined two years of global reporting of climate tipping points (thresholds which if passed could lead to rapid or abrupt climate change) between 1 January 2006 and 31 December 2007 (see Antilla, 2010). The sample included press reports from around the world drawn from English language speaking newspapers and magazines and the BBC News website. The study found that the highest volume of articles in 2006 was published in Europe, especially the UK. In fact, four British media outlets published more articles on climate change feedback loops than all US outlets combined. Somewhat surprisingly, in 2007 media coverage of climate change tipping points fell in the UK, Canada and the US, despite the release of the fourth IPCC Report. However, in all other regions of the globe reporting of feedback processes and abrupt climate change increased, notably in the Asia Pacific region (Antilla, 2010).

A final example is that of research on German media coverage, and the relations between risk communication and discourses on climate change in science and policy. This study undertook discourse analyses of media coverage in 23 publications from 1975 through 1995, including *Der Spiegel*, *Die Frankfurter Allgemeine Zeitung* and *Suddeutsche Zeitung*. The study mapped interactions between science, policy and media, and illustrated the dynamic or 'unstable' and contested discourses unfolding within and between them, thereby influencing public understanding and engagement in climate change action. The authors concluded: 'in the German discourse on climate change, scientists politicized the issue, politicians reduced the scientific complexities and uncertainties to CO_2 emissions targets, and the media ignored the uncertainties and transformed them into a sequence of events leading to catastrophe and requiring immediate action' (Weingart et al., 2000: 280). Overall through examinations in various country contexts such studies have sought to carefully examine the role of the media in climate change science and policy, through empirical examples of key factors and interactions at this interface. Each study has contributed to mapping the contours of interactions between science, policy and practice, via mass media representations.

As mentioned earlier to date most studies focusing on media coverage of climate change have been undertaken in the US and the UK (see Anderson, 2009). However, this is beginning to gradually change as more researchers are undertaking content analyses of reporting in the developing world, as well as non-English language media in industrialised countries (Painter, 2010). In particular, media scholars have begun to focus more upon Latin America, China and India (e.g.

Billett, 2010; Takahashi, 2011; Wu, 2009). One of the largest comparative studies to date was undertaken by James Painter of the Reuters Institute (2010) who examined press coverage of COP-15 in 12 different countries: Australia, Brazil, China, Egypt, Mexico, Nigeria, Russia, UK, US and Vietnam. The authors acknowledge that the study was limited in that it only examined two newspapers per country, but the findings are broadly consistent with those of other studies. The findings suggest that national newspaper coverage of COP-15 focused primarily upon the political drama rather than the science of climate change. This is supported by a more extensive analysis of media coverage of the Copenhagen Summit that was undertaken by MediaClimate. The latter found that 23 per cent of the voices were from 'civil society' (mainly NGOs) while only 15 per cent were from scientists. The Reuters Institute study found that less than 10 per cent of the articles in their sample focused mainly on the climate science and 80 per cent of the articles devoted less than 10 per cent to discussing the science. Moreover, they found that only 12 per cent of the voices quoted were scientists; the main sources were representatives of international and national organisations (including government) such as the Met Office and the Environmental Protection Agency – who both published reports designed to coincide with the event.

Part of the reason for this lack of reporting of the science is that, as James Painter observes, governments and NGOs had many more media relations officers present at the summit compared with universities. Greenpeace, for example, had a team of 20 – more than all of the universities represented there combined. Another possible explanation is that scientists were less likely to put themselves forward to the media because of the intimidating environment in the wake of the hacked UEA emails (Painter, 2010: 11). The study also suggested that there were marked differences in the volume of press coverage that focused on COP-15 in different countries, although they acknowledge that different results may well have emerged if an alternative selection of newspapers were used so we need to view the findings with some caution. The findings suggest that Brazil and India had the highest volume of articles at the start and end of the Copenhagen Summit and Nigeria, Russia and Egypt had the least (Painter, 2010: 40). Indeed, China, India and Brazil were reported to be sending almost 300 journalists to the summit (Vidal, 2009). James Painter concludes:

A gap seems to be emerging between less developed countries and the powerful developing giants like Brazil, India and China, where there is evidence – at least in the print media – to suggest an underlying upwards trend from 2006 until Copenhagen in December 2009. In many poorer countries the coverage remains woefully inadequate and insufficient to match either the scale of the problem or the degree of vulnerability.

(2010: 19–20)

To briefly summarise then, despite an array of methodological issues that make it difficult to be precise about the extent to which media coverage of climate change varies between countries and across continents, it is evident that there are some significant differences that tend to be glossed over in Castells' analysis.

As mentioned earlier Castells argues that celebrities have played a key consciousness-raising role in propelling climate change into the media spotlight. In the next section we turn to consider the evidence about such developments and the extent to which they have fostered public awareness and engagement across the globe.

The role of celebrities in greening the media

As Castells rightly observes over the past few decades celebrities have become increasingly involved in climate change advocacy campaigns and there has been a shift to entertainment-oriented 'soft' news. Growing efforts are now devoted to organising celebrity activities more systematically. Indeed, celebrity involvement in charitable work has become so widespread a form of promotion that it has virtually become part of the job description of an established star (Littler, 2008). Companies such as Celebrity Outreach (www.celebrityoutreach.com) earn their income in part by connecting up celebrities with charitable causes (Brockington, 2009). Similarly, organisations such as the Science and Entertainment Exchange help forge links between scientists, environmentalists and celebrities.

However, it is hard to assess the direct effect of such developments on patterns of press reporting of climate change over time. It is still unclear as to whether celebrity involvement leads to increased prominence of the issues since there are a multitude of factors that play a part in explaining the peaks and troughs of media coverage. In a case study focusing upon celebrities involved in environmental advocacy undertaken by Thrall and colleagues, 165 celebrities were connected to 53 environmental groups (Thrall et al., 2008). However, the analysis suggests that celebrity advocacy had a relatively small impact on environmental news coverage. They claim that large, well-resourced NGOs such as Greenpeace rely very little on celebrity publicity and yet gain extensive news coverage. But they do acknowledge the difficulties with trying to assess the impact of individual celebrities on news media agendas and argue that perhaps it is more appropriate to examine the general influence of celebrity-linked coverage with the rising salience of environmental issues in news agendas. Their analysis suggests that as climate change became more newsworthy it was more likely that celebrities got involved and they gained more news coverage, so they seem to have been *following* rather than leading such developments.

Celebrity endorsements, while often increasing the profile of an issue such as climate change, do not automatically translate into wider cultural acceptance of a political message. Live Earth, the huge benefit concert that took place in 2007, did not generate the same kind of overwhelming response as the earlier Live Aid rock concert that drew attention to the Ethiopian famine of the 1980s. On 7 July 2007 Live Earth began a series of 24-hour worldwide concerts that were broadcast across seven continents (involving 132 countries) through television, radio and internet streams, tuned into by around two billion people. Yet it received very mixed reviews and sections of the UK press were quick to attack it as hypocritical given the size of the associated carbon footprint (Cottle, 2009). A telephone interview survey of US public opinion carried out two weeks following Live Earth found that it appeared to have had limited impact on the attitudes of those who watched the show (Leiserowitz, 2010). However, as the researchers observe, it was not possible to accurately assess the impact of one single event and it is hard to assess whether those individuals who tuned in were already likely to be more supportive of the cause. Over a third (39 per cent) claimed that in the weeks after the rock concert they took individual actions to curb their own greenhouse gas emissions and 20 per cent said that they had sought out further information on climate change. The researchers concluded that Live Earth appeared to have had 'no immediate impact upon American public opinion as a whole' but that it did reinforce/amplify concerns about climate change among individuals who watched the concert.

Also, the involvement of celebrities may soften the message that an environmental NGO seeks to deliver. As Meyer and Gamson argue: 'Celebrities bring with them significant incentives to shift movement frames and in particular to de-politicize or de-radicalize movement claims. Participation by celebrities, then, can speed up the process by institutionalizing or domesticating dissent' (1995: 188).

Whether or not this tends to promote more consensual politics depends, to some extent, on the choice of the individual celebrity and on the medium used to convey the message. As Thrall et al. (2008) observe, an entertainment-oriented form of communication such as Live Earth is focused around entertaining people rather than political persuasion and too much emphasis on political messages runs the risk of alienating people. Indeed, the failure of Live Earth suggests that celebrity endorsement of complex environmental issues such as climate change may distance citizen action through the commodifying practices of celebrity culture. Live Earth may have brought greater visibility to the issues but it seems to have had a very limited impact upon political engagement. As Boykoff et al. argue: 'celebrity efforts may trivialize behavioural change and only serve to ramify contemporary forms of consumerism' (2010: 10).

To briefly summarise then, I have suggested that Castells is right to draw attention to celebrities as a new authorised definer of climate change but his analysis is a little too optimistic about their power to bring about sustained change; while they have given greater visibility to the issues in Western society there is little evidence to suggest that this has resulted in a sea change in mobilisation.

For Castells the key source of power within network society, however, lies in the *switchers* who connect media, business and financial networks (see Chapter 2). In *Communication Power* Rupert Murdoch, CEO of News Corporation, is identified as possessing considerable switching power. In the next section we turn to examine the role Murdoch has played in the climate change debate.

The switching power of Rupert Murdoch

In May 2007 Murdoch made public the company's intention to be carbon neutral, across all their businesses, by 2010. Previously known to be rather sceptical about the reality of climate change, Murdoch's new agenda was thought to have been influenced by the views of his son James. Chief Executive of BSkyB James Murdoch, who had already made BSkyB 'carbon neutral', is said to have persuaded his father to screen Al Gore's film, *An Inconvenient Truth*, at a News Corporation Summit in California during the summer of 2006. A year later, in a high profile News Corporation speech, Rupert Murdoch set out his new vision:

> The climate problem will not be solved without mass participation by the general public in countries around the globe. And that's where

we come in. We're starting with our own carbon footprint... We can set an example, and we can reach our audiences. Our audience's carbon footprint is 10,000 times bigger than ours... That's the carbon footprint we want to conquer. We cannot do it with gimmicks. We need to reach them in a sustained way. To weave this issue into our content – make it dramatic, make it vivid, even sometimes make it fun. We want to inspire people to change their behaviour.... The challenge is to revolutionize the message. For too long, the threats of climate change have been presented as doom and gloom – because the consequences are so serious. We need to do what our company does best: make this issue exciting. Tell the story in a new way... news coverage of this issue is increasing, but we can also do some things that are unexpected... there are limits to how far we can push this issue in our content. Not every hero on television can drive a hybrid car. Often times it just won't fit. We must avoid preaching. And there has to be substance behind the glitz.

He continued:

But if we are genuine, we can change the way the public thinks about these issues. Now there will always be journalists... including some of ours... who are sceptical, which is natural and healthy. But the debate is shifting from whether climate change is really happening to how to solve it. And when so many of the solutions make sense for us as a business, it is clear that we should take action not only as a matter of public responsibility, but because we stand to benefit.

(Murdoch, 2007)

Shortly following the News Corporation Summit in July 2006, to which Al Gore was invited and his *An Inconvenient Truth* documentary screened, a change in position among News Corporation media outlets became evident. For example, in September 2006 *The Sun* (the UK's highest-circulating newspaper) proclaimed in an editorial: 'Too many of us have spent too long in denial over the threat of global warming. The evidence is now irresistible' (11 September 2006). And the front page of this red top newspaper declared: 'Today and every day this week, the *Sun* urges its army of readers to think green.' Two double-page spreads were devoted to *An Inconvenient Truth*, announcing that the film was opening in the UK on 15 September, and a banner appeared across 27 of the newspaper's first pages providing an array of facts about global warming and what actions could be taken.

In 2007 as the issues rose in prominence on the political agenda following the publication of the Stern report and the first volume of the fourth IPCC report, Prince Charles, former UK Prime Minister Tony Blair and comedian Alistair McGowan authored articles in *The Sun* on ways of tackling climate change. With initiatives such as the Global Cool Campaign, the involvement of celebrities was increasingly becoming a feature of popular newspapers' treatment of climate change issues (Anderson, 2011). In 2008 the newspaper partnered with Southern Electric to give away 4.5 million energy-efficient light bulbs, turning half of its front page green for the edition.

However, previously the newspaper's coverage of global warming was thin and sporadic. Between October 2000 and the end of December 2006 *The Sun* and its Sunday counterpart, *The News of the World*, only carried 18 headline stories on global warming (see Gavin, 2007). At the same time between 1997 and 2006 *The Sun* published a number of opinion pieces that ridiculed concerns about climate change (McKnight, 2010).

Since this announcement National Geographic Channel launched the Preserve Our Planet series, MySpace introduced a channel dedicated to climate change at http://www.myspace.com/ourplanet and News Corporation launched its own energy blog. Fox News Channel, a key News Corporation outlet in the USA, launched their 'Green It. Mean it' campaign for Earth Day in 2008 (see http://www.fox.com/greenitmeanit/). However, before 2007 the opinions expressed by the channel's hosts tended to be sceptical of climate change. The news channel often framed the issue by associating concern about climate change with left-wing and liberal politics (McKnight, 2010). A further example is *The New York Post*, another important News Corporation outlet in the US that, following Murdoch's speech, partnered with the Go Green Expo to attract advertising for a special Earth Day section of the newspaper in 2008.

However, more recently Fox News has been accused of enforcing climate scepticism; a leaked email sent in December 2009 by the Washington's managing editor to its journalists in the run up to the Copenhagen conference is said to have asked them to: 'refrain from asserting that the planet has warmed (or cooled) without IMMEDIATELY pointing out that such theories are based upon data that critics have called into question. It is not our place as journalists to assert such notions as facts, especially as this debate intensifies' (Goldenberg, 2010).

In Australia the flagship News Corporation national daily newspaper, *The Australian,* has also been the target for much controversy over its coverage of climate science. In December 2010 it published a full-page feature setting out its position of climate change, claiming that it had been consistently misrepresented particularly through social media:

> There is no dispute that The Australian has opened its news and opinion pages to a wide range of views on the existence and extent of climate change and what should be done about it. But the position taken by the newspaper in its daily editorial column, or leader, has been clear for well more than a decade. In the editorial of April 6, 1995, the paper said: The scientific consensus that global warming is occurring unnaturally, primarily as a result of industrial development and deforestation, is no longer seriously disputed in the world... The Australian has unwaveringly supported global action to combat climate change based on the science.'
>
> (Lloyd, 2010)

An accompanying editorial, 'Truth is Twitter's first Casualty' declared 'Clearheaded readers would find it ridiculous to conclude that The Australian does not believe in climate change... the hothouse of twitter has become a breeding ground for falsehoods that quickly become received wisdom with repeated telling' (Editorial, 4 December 2010). This was in direct response to comments at a conference made by a former journalist that worked on the newspaper, where she expressed a highly critical view of the newspaper's coverage of the topic.

On the basis of an analysis of editorials and opinion columns between 1997 and 2007 a study by McKnight concluded that the newspaper tended to take a generally sceptical line, framing the issue in terms of politics rather than science (McKnight, 2010). He claims: 'While its editorials tended to be sceptical, the newspaper carried a large number of articles and columns which more sharply denied the science of climate change, often drawing on the arguments of a small group of fossil-fuel funded sceptical scientists' (2010: 700). Another study examining the same period concluded that *The Australian* expressed more uncertainty and was more critical of climate science compared with another prominent national daily, *The Age* (Nolan, 2010). An earlier study by McManus (2000) focusing on the Australian press coverage of COP4 (the meeting after Kyoto in Buenos Aires in November 1998) found that *The Australian* newspaper only used headlines with a negative tone in its reporting of the climate conference.

Identifying Rupert Murdoch as a powerful player in shaping climate change coverage should not obscure the fact that explanations for patterns of news reporting are always multi-dimensional. In this case it is likely that a nexus of intricately interwoven interests – business, political and media-related – coincided. In 1997 Rupert Murdoch joined the board of directors of the Cato Institute (a libertarian think tank which has gained a considerable amount of funding from the owner of the biggest private oil companies in the US) (McKnight, 2010: 696). As Newell argues, 'Since the media are controlled by parties with a direct interest in the global warming issue, their assessment of suitable courses of action and the effectiveness of prevailing government strategies is arguably loaded' (2000: 87). A number of climate denial campaigners involved with the Cato Institute were regular guests on Fox News, one of News Corporation's key news outlets in the US. McKnight claims that over the period 1997 to 2007 there were:

> a number of similarities among media outlets owned by the corporation in different countries. It suggests that News Corporation's corporate attitude made it susceptible to, if not ideologically in sympathy with, the advocacy of conservative think tanks which paralleled a campaign by the fossil fuel industry to deny climate change.
>
> (2010: 694)

However, it is important to note that this does not mean to suggest that the various news outlets owned by News Corporation were speaking with the same voice. As McKnight acknowledges: 'In spite of this corporate policy, the intensity of the commentary on climate change was not uniform across the different media outlets of News Corporation' (2010: 703). This should not be surprising since the news values of conflict and controversy, along with the journalistic norm of 'balance', play an important role in influencing how the news media represent environmental issues (Boykoff & Boykoff, 2004; Carvalho, 2007). Moreover, different news outlets are targeted at different audiences and have their own differing ideological standpoints (Carvalho & Burgess, 2005). While mainstream media are likely to steer clear of reporting which severely challenges corporate interests that have a close connection with their outlets, it would be far too simplistic to suggest that ties with fossil fuel industries totally prevent critical reporting. As Carvalho (2007) observes there is always some space for a diverse range of views to be aired. Indeed, investigative reporting by such journalists as Andy Revkin in the US and George Monbiot in the UK have played an important role in exposing the politics of climate science scepticism (see Anderson, 2010).

Castells notes how between 1970 and 2005 there was an explosi
think tanks in the US, often with strong industry links and prioritising
media visibility, yet he makes little mention of climate change sceptic
think tanks. Over recent decades these have played a key role in programming communication networks, especially in the US, and it is to
this that we now turn.

'Climategate' and counter-power

Think tanks and Astroturfing

The growth of digital communication networks has arguably aided climate sceptic think tanks as much as it has helped environmental NGOs
diffuse their messages to publics. A number of studies have highlighted
how powerful industry groups, special interest lobbies and PR companies in the US have manipulated scientific claims and exploited the
news media (see Antilla, 2005). An industry group – Global Climate
Coalition – was established in 1989 to challenge the scientific basis of
global warming. This group, which had 54 industry members, spent
great sums of money on lobbying and public relations. In 1993 alone
one member of this group, the American Petroleum Institute, was said
to have spent $1.8 million on a public relations company to try to defeat
a proposed tax on fossil fuels (Gelbspan, 1995). There have also been
very strong links between the oil giant Exxon Mobil and climate sceptic
think tanks, such as the Washington-based Competitive Enterprise Institute and the American Enterprise Institute (Gelbspan, 2004). The UK's
Royal Society found that in 2005 Exxon Mobil distributed $3.9 million
to 39 organisations challenging the science of global climate change.
Moreover, according to the US Union of Concerned Scientists between
1998 and 2005 the company contributed almost $16 million to a net-
work of 43 groups that questioned the scientific consensus on global
warming.

In September 2006 the Royal Society wrote a letter to Exxon Mobil
demanding that it withdraw its support for climate sceptic think tanks
(Adam, 2006). Jacques et al. (2008) conducted a quantitative analysis of
141 English language environmental sceptical books published between
1972 and 2005 and found that over 92 per cent were connected to conservative think tanks (CTTs), mainly based in the US. Moreover, their
analysis of 50 think tanks revealed that 90 per cent of those that are
concerned with environmental issues support scepticism in their publications and websites, and all of the eight think tanks that are principally
interested in climate change adopt a sceptical viewpoint and are based

in the US. A major strategy of the CTTs is to sow the seed of doubts or in their words 'manufacture uncertainty' (2008: 362). As they argue:

> The central tactic employed by CTTs in the war of ideas is the production of an endless flow of printed material ranging from books to editorials designed for public consumption to policy briefs aimed at policymakers and journalists, combined with frequent appearances by spokespersons on TV and radio.
>
> (Jacques et al., 2008: 355)

An article in the UK's *The Guardian* newspaper revealed that the American Enterprise Institute, a think tank funded by Exxon Mobil, had offered scientists and economists $10,000 each as payment for articles designed to undermine the release of an IPCC report (see Sample, 2007).

Another tactic, 'Astroturfing', is where fake grassroots movements are created to give the impression of spontaneous public protest. For example, the Energy Citizens Group was started by the oil industry (see Johnson, 2009; Lean, 2009). Greenpeace obtained a leaked memo from the Head of the American Petroleum Institute urging members to get involved in a series of anti-climate bill rallies in 20 states across the US. According to the Energy Citizens website:

> Energy Citizens are voicing their concerns about the impact climate legislation passed by the U.S. House of Representatives would have on American jobs, families and businesses. The alliance is urging the Senate to get it right and make sure that climate, energy and tax legislation would not take money out of Americans' pocketbooks and cost millions of jobs. http://energycitizens.org/ accessed 15 October 2009.

Climategate

In November 2009 a major controversy erupted after a large volume of emails, either to or from climate scientists at the UK's University of East Anglia (UEA) Climatic Research Unit (CRU), were made public over the internet. On 17 November a file containing over 1,000 emails was downloaded on the RealClimate website. The timing was significant as this occurred in the run up to the United Nations Conference on Climate Change in Copenhagen (COP-15), which was held from 7–18 December 2009. That the leaked or hacked emails were released in the run up to COP-15 ensured that the story received maximum media coverage (Pearce, 2010). The following day after the file containing the emails was downloaded, Senator James Inhofe, a well-known US climate change sceptic, announced that he was going to be attending COP-15 as a 'One Man Truth Squad' to present a different view.

It is not possible to be completely sure how the journalistic short-hand phrase 'Climategate' was coined but it seems that it originated in the micro blogosphere (i.e. Twitter) (Norton, 2010). The first release of the emails was from a series of bloggers on the West Coast of the US. At 6.20am on 17 November somebody attempted to load a zip file containing the copied CRU emails and other documents from a Turkish Server onto RealClimate – a commentary site on climate science by climate scientists (Pearce, 2010). The post was removed and the CRU alerted. Approximately one hour later someone anonymously posted a link to the data on the Climate Audit blog but the site administrator quickly took it down. Later on that day a link to a Russian server that held the zip file was posted on another conservative blog, Watts Up with That (said to be the most visited climate change blog in the world with over 2,000 unique visitors a month). Further posts soon followed this on The Air Vent and Climate Skeptic, both of which did not have a moderator (Norton, 2010; Pearce, 2010). Very soon after Watts Up with That started using the phrase 'Climategate' (one commentator even recommending that it was used as a framing tool – see Norton, 2010) it got taken up by James Delingpole's *Daily Telegraph* blog with tweeters following suit.

The editorial backlash following Climategate

For many editors and journalists Climategate marked a turning point; it was perceived as a 'game changer', a seminal moment when attitudes among media management were severely jolted. Margot O'Neill, an Australian broadcaster seconded to the Reuters Institute, interviewed 14 mainly UK based journalists and editors (broadcast and print) in May and June 2010. What she found was that many of the reporters whom she interviewed experienced considerable hostility from their editors in the wake of Climategate, particularly after *The Guardian* newspaper launched a detailed investigation into the content of the leaked emails (Pearce, 2010). A senior journalist observed: 'I've never been this hated by our editors' (O'Neill, 2010: 30). Others reported that editors believed that environmental correspondents had got too close to their subject material and had not exercised enough distance. Afterwards a number of editors became convinced that 'climate change was a "scam." One [press] member [was] cold-shouldered by editors and . . . accused of wasting time and resources...We were back at the end of the bulletin, if at all' (Black, 2012). Indeed, following Climategate many journalists claimed they were put under pressure to give sceptical views a greater airing (O'Neill, 2010). Ben Stewart, Head of Media at UK Greenpeace, observed:

> There's a natural pendulum swing in news stories anyway but that happened on stilts with Climategate. It was suddenly like a damn bursting and the media felt like it had to give into an instinct and run what they believe was the 'other side of the story'.
>
> (O'Neill, 2010: 31)

This editorial backlash, however, was not something that was completely new. If we look at the history of environmental reporting in the UK, for example, a similar thing happened in the wake of the furore over the decommissioning of the Brent Spar oil loading and storage buoy in 1995, when there was an editorial backlash against Greenpeace in the UK national media. Previous studies have shown that environmental reporting has experienced many peaks and troughs over time when a substantial elevation in the prominence of the issues has been followed by editors complaining that their reporters had 'gone native' and a general fatigue with the topic (Anderson, 1997: 2010).

Following COP-15 climate change clearly dropped off the news agenda for a while in many countries. O'Neill reported that while there were almost 4,000 journalists who attended the Copenhagen conference, only 150 were present at the follow-up negotiations in Bonn in May 2010 (O'Neill, 2010). In a One World panel discussion in June 2010 the Head of Environment reporting at the UK's *Guardian* newspaper, Damian Carrington, observed 'It's difficult to get editors interested' (cited by O'Neill, 2010: 33). Yet seven months previously on the opening day of the Copenhagen summit *The Guardian* led a special initiative which saw 56 major newspapers in 45 countries join forces to speak with a single voice through a shared editorial urging world leaders to take action. The newspaper's editor-in-chief, Alan Rusbridger, observed: 'Newspapers have never done anything like this before – but they have never had to cover a story like this before' (see Katz, 2009). A notable exception was any leading US newspaper signing up to the joint editorial although *The Guardian* claimed that all but one of the US newspapers approached supported the project. The link below shows which countries joined the editorial http://www.guardian.co.uk/environment/2009/dec/06/papers-copenhagen-leader?intcmp=239

Public opinion in the wake of Climategate

The findings have been mixed on the impact of Climategate upon public opinion. Public opinion on climate change is influenced by a multitude of factors including the state of the economy, views about the government or administration and recent weather patterns; we need to take this into account when assessing the findings of surveys which seek to determine the effects of Climategate on public perceptions. Moreover, question wording can sometimes skew responses – Leiserowitz et al. (2014), for example, found that 'global warming' and 'climate change' mean different things to different people and evoke divergent responses among Americans. Bearing in mind these caveats, and the difficulties of associating cause and effect, a US study that reviewed the findings of a substantial number of studies found strong links between levels of media attention to global warming and trends in public awareness over the past two decades (Nisbet & Myers, 2007). The media are acknowledged to be one of the main sources of uncertainty and confusion about climate science (Downing & Ballantyne, 2008: 17; Nisbet, 2008).

A study undertaken by Jon Krosnick in 2010 is one of the few surveys to have found little evidence of a significant *fall* in the numbers of Americans who said they believed global warming is occurring (Krosnick, 2010). He concluded that Climategate had no major influence, but the telephone survey was undertaken between 17 and 29 November so it is limited to the initial flurry of reporting. By contrast, a study undertaken by Leiserowitz et al. (2010b) concluded that Climategate did have a noteworthy impact upon American attitudes: 'The results demonstrate that Climategate had a significant effect on public belief in global warming and trust in scientists. The loss of trust in scientists, however, was primarily any individuals with a strongly individualistic worldview or politically conservative ideology' (2010b: 1).

Whatever the particular impact that the negative coverage of Climategate had on people's attitudes, it is important to note that a growth in climate scepticism among British and American publics appears to *predate* the UEA controversy. A study undertaken in September 2008 (by researchers at Cardiff University, UK) found that twice as many people agreed that: 'claims that human activities are changing the climate are exaggerated' compared with five years before. Four in ten believed that many leading experts still questioned the evidence, with one in five being identified as 'hard-line sceptics' (Chand, 2009). Moreover, an IPSOS MORI study carried out in the UK between January and March 2010 found that the majority of survey respondents (78 per cent) said they believed that climate change was taking place but this figure was 91 per cent in 2005. Around a third (31 per cent) of the 2010 sample thought climate change was purely caused by human activities. So there appears to have been a small but significant drop in levels of certainty among British publics.

Surveys have painted a similar picture in the US. Moreover, the Gallup 2009 environment survey, undertaken in March of that year, found that 41 per cent of Americans thought the danger of climate change had been exaggerated in the mainstream media, whereas only 28 per cent believed the media had underestimated the dangers of global warming. While there has been a certain degree of volatility since 2001 this is the highest reported level of the 'exaggerated' response since Gallup introduced this measure in 1997 (see Gallup, 2009). Younger people's views stayed the same but those aged 30 and above were more likely to say that news coverage of global warming was exaggerated, and there was a sharp rise among Republicans and Independents. Moreover, the findings suggested that Americans expressed slightly less concern over climate change than in previous years and it was the only environmental issue surveyed where public concern was reported to have dropped.

A Gallup poll conducted in March 2011 found that fewer Americans appeared to be as concerned about global warming compared with three years earlier (Newport, 2011). Moreover, the number of global warming sceptics rose from 11 per cent to 18 per cent. Almost half of the sample (48 per cent) claimed that they believed that the seriousness of global warming was generally exaggerated (a particularly prevalent view among Republicans), 26 per cent that news regarding the issue was generally correct and 29 per cent that reporters tended to underestimate the seriousness of the problem. This represented an increase of around 7 percentage points from 2009 (see Figure 4.6).

The Gallup pollsters conclude that: 'The average American is now less convinced than at any time since 1997 that global warming's effects have already begun or will begin shortly'. Also, the proportion of Americans who thought most scientists believe global warming is occurring fell by 13 points compared with two years previously, the lowest since the first time Gallup included this question in 1997 (see Figure 4.7).

Moreover, an annual Gallup environment poll found that 51 per cent of the sample claimed that they worry a 'fair amount' or a 'great deal' about climate change, representing a drop from 66 per cent in 2008 (Jones, 2011). When asked whether the problem is exaggerated in the news, most (43 per cent) thought it was overstated, 26 per cent

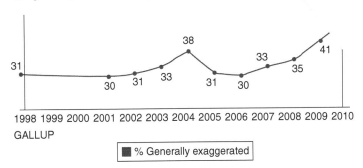

Thinking about what is said in the news, in your view is the global warming – [ROTATED: generally exaggerated, genera... is it generally underestimated]?

GALLUP

■ % Generally exaggerated

Figure 4.6 Views on the seriousness of global warming
Source: http://www.gallup.com/poll/126560/Americans-Global-Warming-Concerns-Continue-Drop.aspx.

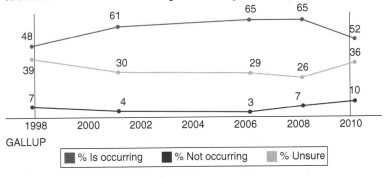

Just your impression, which one of the following statements do you think is most accurate – most scientists believe that global warming is occurring, most scientists believe that global warming is NOT occuring, or most scientists are unsure about whether global warming is occurring or not?

GALLUP

■ % Is occurring   ■ % Not occurring   ▨ % Unsure

Figure 4.7 Perceptions about the scientific consensus on global warming
Source: http://www.gallup.com/poll/126560/Americans-Global-Warming-Concerns-Continue-Drop.aspx

that it was generally correct and 29 per cent that it was generally underestimated (Jones, 2011).

In the US there is a major divide between the Democrats (who mostly accept the risks of climate change) and Republicans (who mostly doubt the risks). Cultural/political identity then is important in understanding how the climate change debate has become so deeply polarised. A longitudinal study by Philo and Happer (2013) found that discussions and news reporting on climate change scenarios tended to strengthen rather than change UK focus group participants' pre-existing positions. Those individuals who had been the least exposed to information were the most likely to be open to reassessing their opinions. Yet understanding public attitudes to climate change is complex and often does not merely reflect a desire to conform to underlying political ideologies/belief systems that are predominant within one's social group. An important difference between the US and the UK is that in terms of attitudes to climate change the former is deeply polarised on right-left party political lines, while the latter is not so clearly divided (Dunlap & McCright, 2008). While political ideology has clearly become a key factor there are multiple 'publics' who may belong to a variety of different networks at any one point in time. Other key intervening factors include age, gender, education and religion. Studies suggest that climate scepticism tends to increase with age and men are more likely to be sceptical compared with women (Pidgeon, 2012). Also, it is important to consider intra-party heterogeneity, given that a variety of competing goals and ideologies exist within a political party and these change over time (see Rolfe-Redding et al., 2011).

Opinion polls on the importance that people attach to the environment versus the economy are mixed. Much depends on the ways in which questions are worded. For example one British study found that in 2010 when asked about the largest problems facing the world (rather than just the UK) people appeared more concerned about climate change than the state of the economy (see Green Alliance Policy Insight, 2012). A recent PEW study of US citizens' public policy priorities found that they routinely ranked global warming at the bottom of the 21 areas. The economy came out consistently top in 2009, 2012 and 2013 (PEW, 2013). Asking people to choose between the environment and the economy presents them as competing values when in reality they are intertwined.

Levels of public concern about climate change have fallen in many countries around the world in recent years (see Ratter et al., 2012). Nevertheless there appears to be a connection between the state of the economy and peoples' attitudes to climate change (Shum, 2012). In the UK during the 2000s (up until 2008) unemployment levels were low and public concerns over climate change were elevated (Philo & Happer, 2013). Similarly there were high levels of public concern in the US in

2007 at a time when unemployment was at its lowest le\
boom years of the Clinton administration (Nisbet, 2011).

If levels of concern have generally fallen over recent year
cism has risen, how are people using the media and whom d
The evidence suggests that when questioned people claim tnat tney are
most likely to trust scientists and environmental organisations, and least
likely to trust politicians and social media/blogs (Donald, 2013; Philo &
Happer, 2013). A recent US study suggests American citizens currently
make little use of social media (or offline media for that matter) to publically communicate about climate change issues (see Figure 4.8). There is
little sense that mass self-communication is being used on a widespread
scale to share and interact about such issues if American findings are
anything to go by.

While the media, especially television, remain an important source of
information for people on climate science we are increasingly saturated
with news. The above findings present a picture of disengagement and
fatigue. As I suggested in Chapter 1 at the current time there appears
to be little evidence to support Castells' observations of a fundamental change in mindset. He claims: 'The bottom line for the media is to
attract audience. The audience gravitates towards news that raises their
emotions. Negative emotions have greater effect than positive ones. And

---

A recent study by Anthony Leiserowitz and colleagues in April 2013 casts some
doubt about the extent to which Americans use online or offline media to
communicate publically about global warming. The study found that:

- Less than 8 per cent of Americans had communicated publicly about global
  warming in the past 12 months (e.g. online or in the media).
- About one in ten Americans wrote letters, emailed or phoned a government
  official about global warming in the past 12 months.
- Of those Americans who contacted a government official in the past 12
  months about global warming, three in four say they urged the official to take
  action to reduce it.
- Only one in three Americans say they discuss global warming at least
  occasionally with friends or family, down eight points from November 2008
- One in four Americans discussed a company's 'irresponsible environmental
  behavior' with friends and/or family in the past 12 months.
- Americans who experienced an extreme weather event in the past year were
  most likely to talk to others about it face-to-face using or by phone, while a
  few did so by social media.

Figure 4.8   How do Americans communicate about climate change?
Source: Leiserowitz et al. (2013: 7).

fear is the most potent negative emotion' (2009: 317). Yet there is also a danger that fear-based portrayals can be counter-productive over time and risk editor backlash.

Summary

The media coverage of climate change is far from linear. Examining trends over time reveals several peaks and troughs which can be related to a number of different factors over and above scientific evidence including: the activities of issue sponsors, policy events, the influence of media owners and news editors, political agendas, the state of the economy, competing news stories, journalistic resources, weather events and perceived levels of public concern. There have been significant variations in the amount of space and level of prominence of climate change, as well as in the way in which the issues themselves have been framed, both in terms of different countries and between different media outlets. Over time the issues themselves have become increasingly politicised and a greater range of social actors have sought to influence the debate.

What of the future? There remain a number of key challenges to be addressed. While climate change has firmly entered mainstream news media agendas, journalists will need to find new ways to sustain attention over time. It is important for the media to provide publics with accurate information about the science of climate change without misrepresenting evidence or exaggerating effects. In the US over 100 climate scientists have now been enlisted by the Climate Science Rapid Response Team to improve the quality of information available to the media and government (see http://www.climaterapidresponse.org/). Also, Climate Central, based in New Jersey, brings scientists and media professionals together in seeking to build a bridge between the scientific community and the public (see http://www.climatecentral.org/). And the Science Media Centre in London is making efforts to increase the number of scientists it is putting forward to the media too (Tollefson, 2010). In the developing world new initiatives are under way to improve training for journalists covering environmental issues including climate change.

However, simply providing people with more and better information does not necessarily directly lead to behaviour change or policy action. An understanding of opportunities and barriers is needed, including an appreciation of intrinsic motivations and infrastructural factors. Also, while political ideology has clearly become an important factor there are multiple 'publics' who may belong to a variety of different networks at any one time. The internet has provided new opportunities for co-ordinating and mobilising action on climate change but we should be wary of approaches that either over-emphasise or underplay its potential power to drive change. Online and offline media interact in complex and dynamic ways.

Further reading

Anderson, A. (2009) 'Media, Politics and Climate Change: Towards a New Research Agenda', *Sociology Compass* 3 (2), 166–82. http://onlinelibrary.wiley.com/doi/10.1111/j.1751-9020.2008.00188.x/abstract

Boyce, T. and J. Lewis (eds.) (2009) *Climate Change and the Media*. Oxford: Peter Lang.

Boykoff, M. (2011) *Who Speaks for Climate? Making Sense of Media Reporting on Climate Change*. Cambridge: Cambridge University Press.

Castells, M. (2009) *Communication Power*. Oxford: Oxford University Press, Chapter 5.

Doyle, J. (2011) *Mediating Climate Change*. Aldershot: Ashgate.

Painter, J. (2013) *Climate Change in the Media: Reporting Risk and Uncertainty*. London: I.B. Tauris.

Philo, G. and Happer, C. (2013) *Communicating Climate Change and Energy Security: New Methods in Understanding Audiences*. London: Routledge.

5
Oil Spills and Crisis Communication

> Never in the history of Spain has an environmental disaster aroused such public outcry, exerted such a political impact, or elicited such media coverage as the Prestige oil spill.
>
> (WWF, 2002 on the Prestige Oil Disaster)

Such was the intense media attention to the Deepwater Horizon oil blowout in the Gulf of Mexico in April 2010 that Tony Hayward, British Petroleum's former harried CEO, famously remarked: 'No one wants this thing over more than I do...I'd like my life back.' Dramatic oil spills make for vivid images of oil-drenched wildlife that can often strongly resonate with emotional attachments to the environment. Large oil spills often fulfil a number of news values including conflict, drama, shock and scandal. However, most oil spills go unreported or fail to attract much media attention. Not all major oil spills receive significant publicity and media coverage is often disproportional to the total amount of damage incurred. It is common for large oil spills to attract a relatively short period of intense nationwide (and sometimes international) coverage, following which it rapidly subsides. However, as we will see in this chapter, this was not the case with the Deepwater Horizon disaster.

In this chapter I examine how the largest oil spill in US history was framed in both mainstream media and the web. I also analyse the tactics used by BP and environmentalists to gain control of the agenda. Over recent decades globalisation and growing reliance upon digital technologies have transformed the news media, intensifying new forms of political activism. Today the media politics of oil spills has to be considered in the context of a rapidly changing global communications environment where many news sources have developed increasingly sophisticated strategies for targeting media and shaping news agendas.

The internet, particularly for activists, is increasingly providing an important co-ordinating tool and a key source of alternative first-hand images and narratives that challenge official accounts.

News values and the framing of oil spills

Oil spills that capture the headlines frequently follow a disaster narrative. Consider the following examples taken from two recent spills:

The Prestige Oil Spill, Spain, 2002

World's Worst Oil Disaster, The Sun, 20 November 2002

A Sunken Timebomb, Daily Mirror, 20 November 2002

BP Oil Spill, April 2010

BP Leak the World's Worst Accidental Oil Spill, The Daily Telegraph, 3 August 2010

Black Death: Will Fisheries Survive the Oil Spill? The Atlantic, 30 April 2010

Compared with routine environmental issues, oil spills tend to have considerable news media appeal. Dramatic pictures of oil-drenched birds, polluted beaches and volunteer rescue teams at work often make for emotive coverage. Photographs or video footage of wildlife – especially seals and dolphins enmeshed in oil – provide vivid images of the trail of destruction that is often left behind. Such disasters possess considerable cultural resonance in terms of deeply rooted ambivalence over environmental protection versus the economy, or the technological domination over nature, versus the notion of industry out of control.

However, there is often little correspondence between the extent of the damage incurred and the level of news media attention. The newsworthiness of particular oil spills is influenced by a number of factors including: closeness to home, socio-economic aspects, and the symbolic and visual dramatisation of the incident (Anderson, 2002). Dramatic oil tanker spills or explosions tend to be much more likely to attract high levels of coverage compared with oil leaking from pipes and storage tanks. Also, the amount of news media attention may be strongly affected by the location in which the accident occurs and its symbolic visual importance. Take the Exxon Valdez (1989) oil spill, for example. It is estimated that 37,000 tonnes of oil was spilt from the vessel – a relatively small amount compared with 223,000 tonnes in the case of the Amoco Cadiz (1978) and 119,0000 tonnes from the Torrey Canyon (1967) (ITOPF). A major reason explaining the intensity of media reporting devoted to the Exxon Valdez oil spill in the United States was that the disaster happened in Prince William Sound. This is a setting that has great symbolic significance for Americans since it is viewed as the ultimate wilderness, the last frontier, while at the same time viewed as an important natural resource to exploit (Birkland & Lawrence, 2002; Wheelwright, 1994; Wilson, 1992).

Since news quickly becomes stale the more unusual or sudden the event the more likely it is to gain novelty value and grab headline attention. While this may increase public awareness of particular risks, the downside of event-centred coverage is that it tends to give readers/audiences the impression that blame can be put down to isolated instances where individuals or corporations have failed in their responsibilities, rather than associated with wider structural issues (Hannigan, 2006). For example, in the case of the Exxon Valdez oil spill coverage tended to be framed around the allegation that the disaster was caused by the drunken state of the Captain, Joseph Hazelwood. This downplayed other possible explanations concerning the oil industry's poor capacity to clean up large spills in areas such as the Prince William Sound, or cuts in funding affecting maritime safety standards (Hannigan, 2006). Indeed, frequently news media frames continue to hold sway long after the event, even when new evidence has emerged concerning wider causes (Darley, 2000).

Oil spills that were once mainly national concerns have increasingly come to be viewed seen as having international importance. Processes of globalisation have led to an elevated a awareness of the transboundary nature of environmental risks (Beck, 1999). Evidence suggests that publics tend to be most concerned about individualised risks which are seen as having concrete, direct impacts, rather than more abstract, distant, threats (MacNaghten, 2006). Over recent decades there has been an intensification of the speed of communication networks and the emergence of 24/7 news coverage has had a significant impact upon the mediation of disasters (Anderson, 2006). In a much more multi-digital, international, interactive and fragmented media environment, there is an increasing need to analyse how different media frame competing rationality claims. To varying degrees publics are reliant upon experts at the national or global level to relay what is happening. The news media increasingly rely upon international news agencies such as Reuters (one of the most heavily accessed news sources on the internet) and Associated Press, to provide rapid information on environmental topics. Alongside this NGOs operate on an increasingly global scale and have become difficult to tell apart from transnational corporations. As I argued in Chapter 3, recent years have witnessed both the increasing proliferation of direct action protests and a rapid growth in the PR industry.

Background to the BP Gulf of Mexico oil spill

On 20 April 2010 a series of explosions occurred on board the BP-licensed Transocean deep-sea oil-drilling rig that killed 11 crewmembers and caused what is widely acknowledged as the largest-ever accidental release of oil into marine waters. It was thought that a sizeable gas bubble entered the well's pipe casing, possibly via gaps in the cement around the wellhead, shooting up to the platform but the blowout preventer failed to activate (Bourne, 2010). The blowout at the well site released over four million barrels (636 million litres) of oil into the delicate ecosystem of the Gulf of Mexico. The oil continued to escape for three months until, following several failed attempts, the blowout was capped in mid-July – a partial success substantially stopping the flow of oil from the ruptured pipeline – with the work finally completed on 19 September 2010. This caused substantial damage to marine and wildlife habitats, and the local fishing and tourist industries. It was clear that BP did not have an adequate contingency response plan in place in the event of such a catastrophe, something that the Oil Pollution Act of 1990 was designed to prevent. Indeed, BP's CEO, Tony Hayward, admitted that: 'it was entirely fair criticism to say BP dropped the ball when it came to planning for a major oil leak' (*Hearings, supra note* 17, at 5 cited in Griggs, 2011). Not only did one of the largest and most profitable companies in the world fail to adequately develop a contingency plan for such an event, it seriously underestimated the magnitude and longevity of the media and public response.

It was only when the rig sunk two days later and oil began to leak that it began to receive much news media attention. Once the impact became visible it started to command attention; images of oiled birds demanded an emotional response from the public. According to one study in the initial week following the accident there was relatively little press coverage, possibly because it was a quite remote location that reporters could not easily access and there was little information to go on (Hoffbauer, 2011). Following this, however, news media attention rapidly intensified over the next two months, particularly as President Obama got involved; his visit to the rig one week after the incident signalled it as an important political issue. Indeed, over the 100 day period following the explosion it received a remarkably high level of attention in the US news media, so much so that it foreshadowed all other news stories – even those related to the economic crisis (PEW, 2010). As Pew (2010) observes its longevity on US news media agendas was extraordinary since most disaster stories peak in the first week and then subside. In the case of Deepwater Horizon the story just kept on going, sustained by regular snippets of breaking news. It was a constantly developing drama with a number of shifting narratives. As the Project for Excellence in Journalism observes:

> That storyline, which included the on-going efforts to cap and clean up the flow of oil and the environmental and commercial damage caused, kept evolving, requiring constant attention and effectively keeping the story from simply defaulting to a more partisan or politicized story line.

(PEW, 2010, un-paginated)

Intense public and news media interest

The oil spill commanded considerable news media attention in the US over an extensive period of time. Between 20 April and 28 July 2010 it made up 22 per cent of the amount of daily space for news, almost twice the amount of coverage that the economic crisis attracted (see Figure 5.1).

It had all the ingredients of a major news story: shock, visual impact, the involvement of elite nations, big business, political drama,

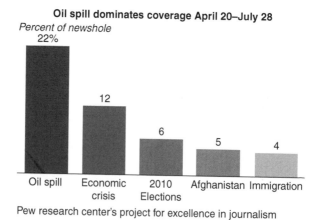

Figure 5.1   Oil spill dominates coverage
Source: http://www.journalism.org/2010/08/25/100-days-gushing-oil/

accusations of cover up, and strong resonance with previous oil disasters such as Exxon Valdez. Moreover, public interest was intense leading some scholars to investigate *reverse agenda-setting effects* (i.e. the question of whether the public rather than the media was setting the agenda (see Ragas et al., 2014). The PEW study, which routinely tracks the degree to which the US public follows news issues, found that often between 50 and 60 per cent of respondents claimed that they were following the story 'very closely' throughout the 100 day period they examined (see Figure 5.2). They concluded: 'If anything, public interest in the Gulf saga may have even exceeded the level of mainstream media coverage' (PEW, 2010, un-paginated). This represents a much higher level of interest than was generated over the healthcare reforms in the US and is even on a par with some of the gloomiest points in the reporting of the economic crisis. When respondents were questioned about which stories

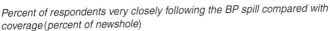

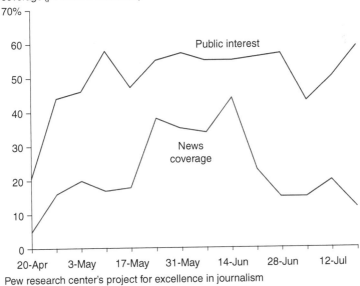

Figure 5.2   Public interest versus coverage
Notes: The dates listed above correspond with the News Coverage Index schedule, which operates Monday through Sunday. The public interest survey was initially conducted Friday through Monday but switched to a Thursday through Sunday schedule as of May 13. The Pew Research Center for the People & the Press is responsible for the News Interest Index.
Source: http://www.journalism.org/2010/08/25/100-days-gushing-oil/

they were following 'very closely' over the time period covered by the PEW survey, the majority ranked the oil disaster as the number one issue in the list for 13 of the 14 weeks (Figure 5.2).

The Project for Excellence in Journalism also questioned respondents about their levels of trust in various sources about the oil leak. As might be expected large numbers of people claimed that they did not trust BP at all. They were also found to be more trusting of news organisations than of federal government (see Table 5.1).

Table 5.1 More trust news organisations than government, BP for oil leak information

Trust information about oil leak from ...	A lot/some %	Not too much/ Not at all %	DK %
News organisations	67	31	2 = 100
Federal government	51	46	2 = 100
BP	39	57	4 = 100

Pew Research Center, 3–6 June 2010. Figures may not add to 100% because of rounding.
Source: http://www.people-press.org/2010/06/09/news-media-trusted-for-information-on-oil-leak/

These findings are corroborated in other studies that found the public (especially younger age groups) viewed the responses of BP and the federal government negatively (Safford et al., 2012). They suggest that government-oriented news media coverage of the spill did not reassure many Gulf Coast residents.

The prominence of the disaster in traditional broadcast media

A comprehensive content analysis of US news media coverage of the Gulf disaster suggests that it was 'first and foremost a television story' (PEW, 2010, un-paginated). It received a great deal of attention on television news, especially cable TV (see Table 5.2). Indeed the story was found to have taken up almost a third of the cable airtime studied (31 per cent) and for CNN, who devoted by far and away the most space to it (mostly covered by veteran disaster reporter Anderson Cooper), this amounted to 42 per cent of its overall airtime. The most prominent focus of US news media coverage, taken as a whole, was found to be on the environmental and economic impact of the disaster and the clean-up efforts. Almost half (47 per cent) of the coverage during the

Table 5.2   Oil spill coverage by media sector

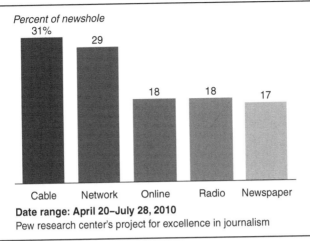

Percent of newshole

Date range: April 20–July 28, 2010
Pew research center's project for excellence in journalism

Source: http://www.journalism.org/files/legacy/Oil%20Spill.pdf

period 10 May–28 July was devoted to the clean up and containment of the spill.

However, cable television was found to concentrate much more on the issue of who was to blame compared with other media outlets.

Considerably less attention appears to have been given to the crisis on US radio, in the national press and online news sites. Radio gave more attention to examining the role of the government compared with any other news outlet – 31 per cent of its coverage was devoted to this aspect.

Coverage in local news outlets

A classic study by Molotch and Lester into the Santa Barbara oil spill of 1969 found that the regional press devoted much more attention to the disaster than the national press (Molotch & Lester, 1975). Thus in terms of frequency-of-occurrence reporting, geographic proximity to the oil spill was a good predictor for the amount of space received. Studies suggest that this pattern was also the case for the Deepwater Horizon oil spill. For example, a cross-cultural comparative newspaper content analysis by Hoffbauer (2011) found that the local newspaper, the *Times-Picayune,* provided nearly twice as many articles mentioning the key term 'BP' during 2010 as the sampled US and UK national newspapers – the *New York Times* and *The Guardian* – and almost five times as many articles as the Canadian *Globe and Mail*. The highest levels of coverage were to be found in the immediate weeks following the blowout; it tailed off once the capping efforts became more successful.

The tendency for the regional press to provide greater space to covering oil spills is borne out in studies of other spills including the Prestige disaster of 2002. Here the local Spanish press provided considerably more sustained and intense coverage of the oil disaster compared with the national press. There were significant differences in framing too. Regional Spanish newspapers focused upon implications for the local economy rather than the effects on wildlife; whereas national newspapers framed the oil spill largely in terms of its environmental impacts and the political controversy over who was to blame (see Anderson & Marhadour, 2007). Nick Jenkins, press officer for the International Fund for Animal Welfare (IFAW), observed: 'local media always take a far higher level of interest and follow the story continuously. National and international media are only interested when the story is at its peak(s)' (Personal communication, 22 July 2004).

As with the Prestige accident, in the case of the BP oil spill the federal government was slow to respond before the environmental effects became evident. The White House's peak in terms of engagement was in the week 14–20 June (when oil executives from BP, ExxonMobil and Shell were testifying in front of the House Sub-Committee on Energy and the Environment), a month later than when the newspaper interest peaked. However, on this occasion it was the national *New York Times* that appears to have initially led the coverage:

> the leading national newspaper the *New York Times* broke the story with the local newspaper only providing limited coverage in the early stages of the disaster. As engagement increased in the leading publication both local and international publications began to follow their lead and published an increasing number of articles. It appears that once the issue built-up, gained momentum, and was signalled as a significant issue, the local newspaper allotted more resources to covering the story and eventually surpassed the other publications' coverage.

> (Hoffbauer, 2011: 42)

In contrast to earlier major oil spills such as the Torrey Canyon or the Amoco Cadiz, the digital revolution has provided a space in which alternative framings can flourish but to what extent did the internet come into its own in the case of the Deepwater Horizon disaster?

Mediation of the oil spill on the web

PEW's small-scale study of online coverage looking at social media found that it only featured in the top five stories in blogs for five of the 14 weeks and its presence on Twitter was even less prominent (see Table 5.3).

However, one content analysis study undertaken by Muralidharan et al. (2011) suggests that BP made much more use of Twitter and Flickr compared with Facebook and YouTube – possibly because of the ease of quickly uploading information using these channels. The discussion of the crisis on social media, as might be expected, appears to have been much more sceptical and negative in tone compared with the traditional media. Muralidharan et al. concluded:

> Based on dominant audience issues on Facebook (personal attacks) and YouTube (boycott BP) coupled with negative emotions, it appears that BP's corrective action did not sway all audiences, nor did it materially affect general public opinion at the time.
>
> (2011: 231)

Table 5.3 Number of weeks as a top five story

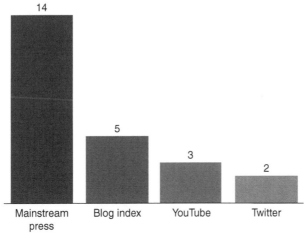

Date range: April 19–July 28, 2010
Pew research center's project for excellence in journalism

Source: http://www.journalism.org/files/legacy/Oil%20Spill.pdf

The PEW study undertook a small-scale study of websites linked to 14 major national broadcast and print news outlets. Compared with earlier major oil spill disasters there was much more scope for mainstream media to utilise sophisticated interactive multimedia features. For example, the *New York Times* used a video animation that enabled viewers to see how a last minute attempt to prevent the spill failed to work. CNN, in particular, was found to stand out in terms of the amount of web tools and features that it used. For example, it provided key figures tracking the amount of oil spilled in comparison with the Exxon Valdez spill, the amount of money spent in BP's disaster response as compared with its annual profit and how different wildlife were affected (CNN, 2010). The websites of many newspapers and television stations had oil spill trackers that graphically demonstrated how the oil slicks moved over time and PBS News Hour used an oil leak widget to track daily the amount of gallons leaked (PBS, 2010). Despite the PEJ finding that blogs played a limited role in covering the Gulf oil disaster some commentators claim that they really came into their own during the crisis:

> A Google search of the gulf oil spill returns 6,865,009 blog entries referencing the event. It is in social media tools that some of the best instant reporting and editorialization is happening. Whether it is Rising Tide North America using their Twitter account to warn volunteers that migrating dolphins will travel through the spill zone, or ethical fashion blogger Joshua Katcher – The Discerning Brute, encouraging his fans to challenge the 'glitz' of BP's heavily funded PR agency and internalize a bitter truth that our relationship with nature is an abusive one.
>
> (Stephanian, 2010)

Views are mixed then about the role of social media in reporting on the accident. As we shall see later on environmental NGOs made considerable use of new communication technologies to question the official version of events and offer alternative accounts, as well as mobilise supporters to dislodge dominant discourses associated with the oil industry. Alongside traditional media the internet also proved to be an important mobilising tool in the Prestige oil disaster. Nick Jenkins, press officer for IFAW, observed:

> The internet is increasingly becoming a major tool in campaigning and that will continue. It has provided a means of reaching large numbers of people without relying on the media. Hence our message

can go out unadulterated. It is still, however, far less important than media, particularly TV.

(Personal communication, July 2004)

Compared with earlier major oil spills, such as the Amoco Cadiz, during the Prestige crisis individuals with internet access could survey and check information quickly. Environmental groups could better co-ordinate action and in December 2002 it was estimated that approximately 200,000 people took to the streets in North West Galicia to protest about the handling of the oil spill (see Anderson & Marhadour, 2007). However, in 2001 Galicia was one of the Spanish regions having minimum internet reach with only around 15 per cent of the population (Salaverría, 2002). Despite the limited internet access in the immediate area affected by the oil spill, its role in terms of co-ordinating more widespread protest action among environmentalists is likely to have been significant. Also, as I argued in Chapter 3, online media have an agenda-setting effect on offline media and vice-versa. The picture painted here then is one of dynamic flows between 'old' and 'new' media rather than one having replaced the other. Further layers of complexity can be identified when we look at how the oil spill was variously framed, to which we turn to examine in the next section.

News media framing of the Deepwater Horizon oil spill

As I suggested in Chapter 3, we can often identify multiple framings of the same event within and between news outlets. While some news outlets placed more emphasis on the ecological destruction others concentrated on the issue of who was to blame. Studies of newspaper coverage of Deepwater Horizon have, to date, concentrated upon a few elite agenda-setting newspapers such as the *New York Times* and less attention has been paid to the popular, more widely circulating, press (see for example, Hoffbauer, 2011; Schultz et al., 2012). A study undertaken by Schultz et al. (2012) suggests that the opinion-leading US newspapers partially adopted BP's frames, whereas the sample of UK newspapers (*The Times, The Sunday Times* and the *Financial Times*) focussed much more on BP being seen as responsible for the solution (2012: 104).

BP issued an increasing number of press releases over the course of the crisis. Before the oil spill BP issued no more than four press releases in a given week, whereas following the sinking of the Deepwater Horizon it started to issue a growing number of press releases to a total of 171 (Hoffbauer, 2011). There have been a number of content analysis studies of BP's press releases drawn from the company's website (e.g. Choi, 2012; Harlow et al., 2011; Schultz et al., 2012). To access BP's website on 22 April 2010 see the following link: http://web.archive.org/web/20100422033150/http://www.bp.com/bodycopyarticle.do?categoryId=1&contentId=7052055

The above studies suggest that BP's press releases tended to concentrate upon what it was doing to clean up the damage and compensate the victims, but there was some attempt made to disassociate itself from being to blame. Wickman notes a 'rhetorical manoeuvring in the company's press releases that directions attention towards actions that followed from the explosion on board Deepwater Horizon but not the deeper systemic problems that led to off-shore drilling in the Gulf or the issues that may arise from BP's on-going presence there' (2014: 14). Studies suggest BP paid little attention to the causes of the disaster and focussed on the solutions rather than the consequences of the oil spill. Tony Hayward, British Petroleum's former CEO, stated on Sky News on 15 May 2010: 'The impact will be very, very modest...the ocean is very big and the oil we're putting into it is tiny.' BP was found to employ what Harlow et al. (2011) refer to as a *decoupling strategy* whereby it sought, particularly in the initial phases, to disassociate itself from responsibility for the cause but at the same time present itself as part of the solution.

Which voices got the most favourable treatment in the BP case? Molotch and Lester (1975) found that the national newspapers tended to rely on national sources and reported on events that could be framed as 'national events', while local coverage carried information that was considered important to people in the local Santa Barbara area. Similarly research on the Prestige disaster highlights how the local press tends to focus on more proximate issues and government sources dominated much of the national coverage (Anderson & Marhadour, 2007; Jiménez et al., 2004).

The PEW centre research found that President Obama was the key primary source in the national media's coverage of the Deepwater Horizon spill, making up 12 per cent of the coverage, with Tony Hayward gaining 4 per cent. This finding is confirmed by other research. For example, Goidel et al. (2012) undertook a quantitative content analysis of coverage of the BP oil spill over the period 20 April to 16 July 2010 by 14 different local television news stations in five different Gulf Coast states (Texas, Louisiana, Mississippi, Alabama and Florida). These represented a mixture of metropolitan areas directly affected by the spill and all three broadcast networks. The study revealed that President Obama was the most prominent news source mentioned in 7.6 per cent of all sampled stories and just under a third of news stories (30.5 per cent) that included an official source. However, other prominent voices included local officials Governor Bobby Jindal and Governor Haley Barbour. They also found important regional differences: Louisiana news coverage emphasised the effects on seafood and the fishing industry while also giving a pro-drilling and anti-moratorium slant to their coverage, while Florida news coverage placed more emphasis on the impact on tourism and an anti-drilling and pro-moratorium slant.

In the case of the Prestige disaster, federal government, congress, the oil industry and state politicians were found to gain the most access to national newspapers not simply in terms of frequency of coverage but also in terms of the much greater likelihood of their views being prominently placed on the front pages (see Anderson & Marhadour, 2007).

News blackouts and information management

High profile oil disasters tend to be heavily managed; the authorities control who can access the area and what they can see. In the case of the Prestige accident it was claimed that broadcasters were told to avoid using the term 'oil spill':

> The central government spokesman's press conferences were always denied in less than 12 hours, by the facts announced in some private TV channels or by Portugal and France. It appears that the government, because of the impossibility of controlling the situation, chose to hide the information by deceitfully reassuring messages, developing obstacles for the media, lies, and even censorship (forbidding for example the flights over the sinking zone). This kind of system doesn't work when it is perfectly evident what's going on and people can contrast it with false official information. The government even prohibited the public TV workers from using the term 'oil spill.'
>
> (Marine scientist, Vilas Paz, 2004)

Indeed, such was the intensity of concern over censorship allegations that a petition, demanding further investigation was presented to the European Parliament in December 2002.

The authorities also tightly controlled mainstream media coverage of the Deepwater Horizon oil spill. Despite the claim that 'Neither BP nor the U.S. Coast Guard, who are responding to the spill, have any rules in place that would prohibit media access to impacted areas' (cited by Ciarallo, 2010), press access to the area appears to have been highly managed by BP and the US Coast Guard. The official US 'Restore the Gulf' website stated in its information for members of the press:

> Media access is a critical component of the overall external communications effort. Media embeds with responders or media embarks aboard response assets are highly encouraged, to the extent they can be safely accommodated. They should be coordinated through the Unified Area Command Joint Information Center (UAC JIC), which will ensure proper credentialing and provide any necessary ground rules or hold harmless agreements. http://www.restorethegulf.gov/category/media-access

BP and the US Coast Guard ensured that they escorted photographers and journalists so that they maintained control over when they could gain access and what they saw. Flight restrictions were imposed on the area, on 'safety grounds', limiting the extent to which private aircraft could fly over the Louisiana coastline (Phillips, 2010). Aircraft that did not receive permission were required to fly 3,000 feet above the restricted area that allowed very little visibility (Phillips, 2010). Any wilful violators of the 20-metre 'safety zone' whether members of the media or public clean-up volunteers, could be fined up to $40,000 or charged with a Class D felony.

There were reports that BP workers had to sign a gagging order preventing them from speaking to the media while working on the clean up (see Frohne & Dearing, 2010). Initially contractors were forbidden to speak about the disaster or the clean-up efforts without prior approval under the Master Charter Agreements for BP's Vessels of Opportunity programme (Lyndon, 2012). It also seems that BP blocked members of the media from filming the impact of the disaster, such as a CBS news crew who were said to have been threatened with arrest when they tried to film an oil-covered beach in Louisiana (see Blizzard, 2010; Ciarallo, 2010; Huffington Post, 2010). The Society of Environmental Journalists sent a letter on 4 June to Admiral Thad Allen, former National Incident Commander, expressing their concerns about the restrictions imposed on the media and stating they were: 'deeply disturbed at the growing number of reports we have received that journalists are being prevented from doing their Constitutionally protected jobs; to provide information to the public about the mammoth oil disaster playing out on the

Gulf Coast' (SEJ, 2010). Also, the American Civil Liberties Association (ACLA) of Louisiana sent an open letter to the sheriffs of all the local parishes on 28 June, reminding them to respect freedom of access to information and public assembly rights enshrined in the First Amendment (Juhasz, 2011). Finally in July 2010 the Coast Guard responded to mounting criticism and amended its rules to accommodate journalists in the Boom Safety Zones if they complied with certain rules (Lyndon, 2012).

Bloggers also revealed that BP deliberately doctored some official photographs on its website showing how it was responding to the spill (see Hough, 2010; Korosec, 2010). On 21 July BP admitted to posting a photograph on its website that exaggerated the activity at its command centre (Aravosis, 2011). The photo had shown ten video screens being monitored by its staff but in reality three of them were blank (Lyndon, 2012). This lack of transparency suggested to many that all BP was concerned about was protecting its public image. As Freeman explains:

> BP tried to present the story on its terms, privatizing public information and denying the government custody of information. BP officials claimed the spill information was stock market sensitive, meaning it had to be managed by disclosure rules from the London and New York Stock Exchange. The stock exchange disclosure rules led to a lack of transparency from BP and left the public uninformed on the recovery process.
>
> (Freeman, no date)

Following a hearing in Congress, where BP was accused of withholding information, the company released the first underwater high definition video footage of the oil leak on 13 May via a short video on YouTube, a month after the disaster occurred. It was forced to stream live webcam footage from the ocean floor. It made great television news and was said at the time to be one of the most popular links on CBSNews.com (see Guthrie, 2010). Given these issues of openness and transparency, in the next section we examine BP's media strategy and how this changed over time.

Damage limitation and image repair

Following an environmental accident companies often have to deal with a crisis in public trust and potentially face a legal battle and suffer economic losses. From the start BP sought to distance itself from the oil spill and downplay the extent of damage incurred. According to 'Image Repair theory', organisations attempt to correct negative public perceptions following a crisis via damage limitation exercises (see Benoit, 1995; Harlow et al., 2011). This is known as crisis management in the public relations literature. Image repair strategies often change over time; in the very early days, due to the often-sudden nature of the event, they tend to be relatively ad hoc. In the case of the Exxon Valdez disaster (1989) the company faced an onslaught of unfavourable media coverage and sought to counter this by issuing an 'Open Letter to the Public', playing down the damage to the environment and laying the blame on the captain who had been found to have been drinking before the accident occurred (see Benoit, 1995).

Likewise in the case of the Deepwater Horizon disaster BP invested heavily in using mainstream media to try and get its message across. On 26 May it placed full-page advertisements in major US newspapers including the *Washington Post*, the *New York Times*, the *Wall Street Journal* and *USA Today* explaining what action it was taking to address the crisis. At the end of May the company took on Anne Kolton, former head of public affairs at the US Department of Energy and former spokesperson for Dick Cheney, as its head of US media relations and they formed a new division to oversee the company's response. Shortly after, it launched a television advertising campaign on 4 June featuring former CEO Tony Hayward apologising to viewers saying: 'we will make this right'. This TV campaign was reported to have cost in the region of $50 million (Smith, 2010). Prior to this the corporation spent in the region of $36 million on its 'Helios Power' media campaign in April 2007 (targeting newspapers, TV, internet and radio) that presented BP as an environmentally responsible brand (see Smerecnik & Renegar, 2010). The danger of course with such exercises is that they may be interpreted by the public as insincere.

The company was relatively slow to exploit digital media. However, at the beginning of June BP purchased key search engine optimisation terms on Google, Yahoo and Bing so that its Gulf of Mexico response page would come up as the first sponsored link when someone inputted words such as 'oil spill' (see Morgan, 2010; Fiscal Times, 2010; This Week, 2010). As Figure 5.3 shows, searches for terms such as 'gulf oil', 'spill', 'oil clean up' resulted in BP's sponsored link appearing at the top, with the tagline 'Learn about BP's progress on the oil spill cleaning efforts' (This Week, 2010).

Figure 5.3   BP buys up search engine optimisation terms
Source: Copyright, Lee (2010) 'BP, Crisis Communications and Social Media', 1 July. http://
www.bruceclay.com/blog/2010/07/bp-crisis-communications-and-social-media/

From the end of May 2010 BP was estimated to have top visibility for over 1,000 search terms related to 'oil spill' on search engines and its videos were promoted to the top of search results for videos on YouTube (Allen, 2010). An internal document obtained from *Advertising Age* suggested that the corporation went from spending around $57,000 a month on search engine advertising to spending almost $3.6 million just for the month of June (Learmonth, 2010). The director of a search marketing company, Scott Slatin, claimed that in April 2010 the number of searches on Google for 'oil spill' was 2,240,000 compared with a yearly average of 301,000. He estimated that BP was spending in the region of $7,500 a day to secure a top position in the search hierarchy on Google and a further $3,000 a day for search engine optimisation on Yahoo and Bing (see Fiscal Times, 2010). One estimate puts these search terms as driving 47 per cent of all traffic to BP.com and 22 per cent of all search traffic (see Wickman, 2013). According to the US House Energy and Commerce Committee BP was thought to have spent over $93 million on newspaper advertisements and TV spots in the weeks following the Deepwater Horizon oil spill; this was three times as much as during the same time the previous year (Tracy, 2010).

BP's use of social media: Too little too late?

BP was slow to use social media; it took seven days to send out a tweet (in capitals) with a link to a press release.

The first tweet from BP on May 27th

"BP PLEDGES FULL SUPPORT FOR DEEPWATER HORIZON"

This came 11 days after a satirical tweet was issued from an anonymous parody account posing as the voice of BP:

"We regretfully admit that something has happened off of the Gulf Coast. More to come"

16th May 2010

The first tweet from @BPGlobalPR, the BP parody account

Alongside this strategy BP also updated people via its Facebook page; the first update appeared a couple of days after the first tweet about the incident. Its YouTube page was not started until 18 May, nearly a month after the oil rig exploded (O'Reilly, 2010). They were slow then in utilising new digital technologies and invested most of their efforts on getting their message across through mainstream media. However, as the crisis developed they did exploit the participatory potential of new media more and they employed crowd-sourcing techniques to help them find a solution to the problem. When they asked citizens to submit ideas about stopping the flow and cleaning up the spill over 120,000 suggestions were received (Lyndon, 2012; Restore the Gulf, 2010). The involvement of a celebrity also added additional news interest when the Hollywood actor Kevin Costner volunteered his technical assistance. The company spent £10 million on purchasing his oil spill clean-up machine although it was reported to suffer from several issues.

Although BP used a variety of social media channels it engaged in a largely 'old media' approach, applying classic communications approaches to a new media environment which demanded different tactics. It produced what seemed akin to television advertisements on

YouTube and placed links to technical updates on Twitter. Also, it frequently turned off the comment function on social media outlets so it became a largely one-way channel of communication (Capstick, 2010). Despite having greater information subsidies the corporation failed to win back support from the public. A number of communication gaffes made by its leaders only served to reinforce the view that it was primarily concerned with protecting its reputation rather than engaging in transparent dialogue.

Challenger tactics

Although BP attempted to control information those affected by the oil spill could potentially access a myriad of conflicting accounts. Liu (2010) reports that in July 2010 using the search term 'BP oil spill' produced over 33 million results, more than 500 Facebook groups, over 134,000 YouTube videos and in excess of 1.6 million blog posts relating to the accident. Environmental groups made extensive use of social media and mobile apps (e.g. the Oil Reporter app developed for Crisis Commons) during the Deepwater Horizon disaster to support the recovery efforts. However, blogs appeared to be one of the most heavily used outlets. Melissa Merry (2014) examined the media activities of 13 national US environmental organisations in relation to the spill. The following table suggests that more traditional methods of communication, such as emails and press releases, were less frequently used (see Table 5.4).

Environmentalists proved to be much more adept at using what Castells refers to as 'mass self-communication' (see Chapter 2). As of 1 June 2010 the 'Boycott BP' Facebook page had 233,000 fans while the official BP pages only attracted 18,000. The parody Twitter account

Table 5.4   Distribution of communications in the sample

Communication forum	Number of statements
Blog	787
Email	252
Press release	33
Testimony	22
Total	1,394

Source: Merry, M. (2014), p. 15.

@BPGlobalPR had 125,000 followers while the official site @BP_America only had 10,000 (Andersson & Macdonald, 2010). Fake Twitter feeds continued to appear such as:

> We SAID we'd clean up the Gulf coast. What more do you want from us?
>
> Today is the one year anniversary of something happening off the Gulf of Mexico. Looking into it. #bpcares

The author of the parody account, @BPGlobalPR, wrote an 'Open Letter to the Media' that attracted more publicity (Street Giant, 2010). An excerpt from the open letter to the media

> "I started @BPGlobalPR, because the oil spill had been going on for almost a month and all BP had to offer were bullshit PR statements. No solutions, no urgency, no sincerity, no nothing. That's why I decided to relate to the public for them. I started off just making jokes at their expense with a few friends, but now it has turned into something of a movement. As I write this, we have 100,000 followers and counting. People are sharing billboards, music, graphic art, videos and most importantly, information." (2nd June, 2010)

Environmentalists used a mix of old and new media in their campaigns to highlight the damage caused by the disaster. Theirs was a more interactive approach that relied heavily on symbolic statements and parody. This approach to generating news media attention has a long history. For example, during the Amoco Cadiz oil spill (1978) FoE campaigners left a dead oiled bird on a desk in an office of the directorate of Shell in Paris. Similarly, Greenpeace activists dressed in bird costumes drenched with black 'oil' at a meeting of European ministers discussing maritime safety in the wake of the Prestige. The activists dumped barrels of oil residue outside the EU Headquarters and displayed banners demanding that ministers 'Act Now' (see Greenpeace, 2002). Environmental groups are still heavily reliant on creating 'stunts' to draw the attention of the mainstream media (Hutchins & Lester, 2006). However, campaigning methods have evolved since the late 1970s to increasingly involve media strategies involving sophisticated multimedia techniques.

In response to the Prestige disaster a number of websites were developed to counter official versions of the event and keep up pressure on the authorities. Websites such as www.chapapote.org exposed organisational failures through humour and graphics. Another website, Marea, provided an alternative account of the Prestige oil spill through: 'reconstructing it as a theatrical event in which the different social and ethical archetypes of modern society are played out against each other'. An alternative documentary was produced by Iratxe Jaio and Klaas van Gorkum that aimed to offer a critical, reflexive account of the oil spill (Marea, 2005). They maintain:

> From the start of our production, we have been documenting the visual style of these messages, by taking hundreds of stills from the slogans, banners, posters and graffiti that have been used in the different campaigns of protest we encountered in public space. Typical of these signs is their subversive tone: the street belongs to everyone, and its language presents an alternative to what is on display in the mainstream media. As such, it challenges the establishment, which is in itself an abstraction of the original Prestige story.
>
> (Marea, 2005)

Drama and theatrical techniques were also central to environmental campaigners' media toolboxes during the Deepwater Horizon disaster. On 20 May 2010 Greenpeace launched their 'Boycott BP' campaign. Overnight they projected the words 'Your Logo Here' onto huge fuel storage tanks in the refinery that supplies BP in the South East of the UK (Fernandez, 2010). The next day two protestors scaled BP's Headquarters in London and hoisted a flag with the slogan 'British Polluters', showing their logo dripping in oil.

Using these dramatic means Greenpeace launched their rebrand the BP logo competition. In order to maximise exposure they brought in the actor and television personality Stephen Fry to tweet the logo redesign competition to his (at the time) 1.53 million followers. Greenpeace asked people to submit their entries via a photo group on Flickr, the online photo management and sharing app. This was a clever strategy designed to ensure longevity since, as various commentators have observed, the Flickr groups will long outlive the competition and the images themselves have been widely reproduced on numerous other websites. It was estimated that when the competition closed there were around 2,500 entries and in the region of 600,000 page views of the rebranded logos (Telofski, 2010).

Greenpeace employed a multi-pronged approach and a carefully targeted advert on 20 May was placed in *The Guardian* newspaper (the national UK newspaper that tends to carry the highest level of stories about environmental issues) (see Figure 5.4), inviting readers to enter the logo redesign competition.

Figure 5.4   Greenpeace advert in *The Guardian*, 20 May 2010

In June and July protestors engaged in further publicity stunts at BP petrol stations across the US and the UK, closing pumps and encouraging consumers to boycott the company. Rallies took place in New Orleans publicised through social networking sites such as Twitter and a Facebook group, BP Oil Flood Protest (see Figure 5.5).

Flash mobs took to the streets with activists dressed in black. On 1 June the Mozilla Firefox 'black oil' plugin was released; all mentions of BP on the web could now be covered in 'black oil' as a visual signifier of 'dirty oil'. The famous graffiti/street artist Banksy

Figure 5.5   Dead pelicans parade, New Orleans
Source: Flickr (http://www.flickr.com/photos/infrogmation/4675096078/sizes/m/in/set-7215 7622590579172/)

transformed a dolphin-shaped children's ride into an anti-BP statement and London artists organised a campaign against the Tate Britain to pressure them to refuse BP's sponsorship. Liberate Tate hung dead fish and birds from dozens of giant black helium balloons inside a gallery in a protest against the spill (see Lyndon, 2012).

Environmentalists relied upon celebrities to amplify their voices on an increasingly crowded media stage. As mentioned Stephen Fry helped their logo redesign competition reach a wider audience than they could otherwise have drawn. Celebrities including Sandra Bullock, Drew Brees and Eli Manning were featured in a 'Be the One' video in July, urging people to sign a petition to demand that more money be spent on restoring Gulf Coast ecosystems. A number of musicians including Lady Gaga and The Backstreet Boys announced that they planned to boycott BP on their national tours (Lyndon, 2012). As mentioned in Chapter 2, the involvement of celebrities can bring greater visibility to a cause but it can also act as a double-edged sword; rarely does it fundamentally unsettle established orders.

Associated Press estimated that at the end of the protests sales of BP fuel had declined by between 10 per cent and 40 per cent at some stations. However, it is very difficult to put a precise figure on this since BP does not own many of the stations selling their fuel. The boycott was short lived. A PR week/One Poll survey of public opinion one year on found that 93 per cent of respondents thought the oil spill had been damaging to BP's reputation, but the vast majority (86 per cent) said they had not boycotted any of its products. This raises the issue of 'clicktivism', sometimes referred to as 'slacktivisim', that I discussed in Chapter 2. While social media certainly helped to raise visibility of the issues and question the official version of events during the Deepwater Horizon disaster, what effects did this have on levels of activism? In the final section we turn to consider the growth of grassroots activism and crowd-sourcing enhanced by the digital revolution.

Grassroots mapping and crowd-sourcing

Aided by the digital revolution community-led information sharing and mapping techniques that challenge official accounts have grown rapidly in recent years. However, community-mapping exercises are by no means new and there is a long history of local people documenting environmental abuses (see McCormick, 2012). The Louisiana Bucket Brigade (LABB), a New Orleans-based environmental activist group, first began monitoring local air quality in 2000, using an EPA-approved, community-friendly air quality-monitoring tool. However, more and more citizens now have the technology to be able to participate in disaster responses through, for example, the use of smartphones and social networking sites.

During the initial weeks there was a lack of publicly available high-resolution detailed images of the effects of the spill. A group of US citizens led by Jeffrey Warren of grassrootsmapping.org developed a set of DIY tools in 2010 for sending inexpensive digital cameras up in helium balloons and kites to generate aerial photos of the Gulf of Mexico oil spill. See Figure 5.6.

The cameras were set initially to take continuous pictures once every second. The volunteers anchored the camera to a rig (a soda bottle) and tied it to a balloon. Although traditional aircraft were banned from flying lower than 4,000 feet above the spill the restrictions did not affect tethered kites or balloons (see Figure 5.7).

They then released these images into the public domain so they could be downloaded and republished without permission and created the Oil Spill Crisis Map, whereby citizens could submit health, environmental and economic reports through a variety of platforms, including mobile

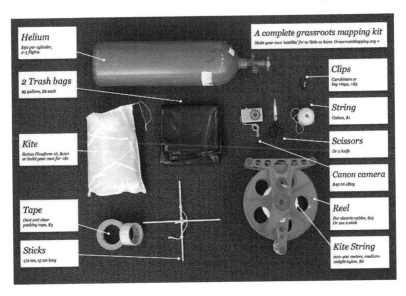

Figure 5.6   Satellite poster 2010
Source: Satellite poster 2010. Grassroots Mapping Community. Photograph, Jeffrey Warren. Creative Commons Attribution License.

Figure 5.7   Grassroots Mapping Community
Source: Grassroots Mapping Community. Photograph, Jeffrey Warren. Creative Commons
Attribution License

phones and the web. The images circulated widely and were used by many mainstream media outlets including the *Boston Globe, New York Times* and many news agencies. Some of the images can be seen here: http://insideclimatenews.org/slideshow/story-grassroots-mapping

The initiative relied heavily upon collaboration through crowd-sourcing to raise funds to contribute to the costs (mainly travel) incurred. The website Kickstarter.com was used as a means by which the group could encourage individuals and organisations to financially 'back' the cause, receiving rewards for different levels of support. For $50 they offered to put backers' names on a balloon and for $1,000 they promised to send backers a complete Grassroots Mapping Kit. By 21 June 2010 a total of 145 backers had supported the project raising $8,285 to support the work. The Grassroots Mapping mailing list and wiki played a key part in helping to co-ordinate volunteers and over 11,000 images were produced. They also collaborated with other groups including Blue Seals, Greenpeace, Louisiana Environmental Action Network and Americorps (Warren, 2010).

The LABB also made use of the Ushahidi platform, open source, web-based software crisis reporting tool for crowd-sourced information gathering (see Warren, 2010). Ushahidi was used by a number of organisations at the time of the Haitian earthquake; it means 'testimony' in Swahili (McCormick, 2012). According to Warren:

> The Ushahidi platform has emerged as a common and easy-to-install system for crowdsourced crisis reporting. Developed in collaboration with Kenyan programmers to help voters report election violence in Kenya in 2008, it allows citizens with mobile phones to send 140-character text messages to a publicized telephone number.
>
> (2010: 31)

However, the sorts of information it provides can be hard to verify, most reports are only a few words without a name or any photographic evidence and data can be falsified. For these reasons Warren acknowledges that it is more of a tool for emergency rather than evidence-based reporting. Nevertheless he contends: 'A combined strategy involving the use of an Ushahidi-like platform with aerial imaging can result in clusters of crowd sourced reports providing target sites for follow up mapmaking sessions' (2010: 32).

Another non-profit organisation **Skytruth** that specialises in analysing satellite images, remote sensing data and aerial photographs enabled official estimates of rate of oil leakage to be challenged. The group claims it played a key watchdog role in the unfolding of the Gulf of Mexico disaster:

> One week after the accident, it raised concerns that the amount of oil spilling into the Gulf was likely much higher than the 1,000 barrels-a-day estimated by BP and government officials. *The New York Times* and other media outlets picked up the analysis published on the SkyTruth blog on April 27. The next day, government officials publicly broke ranks with BP and raised its estimate to 5,000 barrels a day, the amount that SkyTruth had calculated.

SkyTruth's Oil Spill Tracker site enabled Gulf residents to act as citizen journalists uploading their own photos, videos and commentary. The group claims that in the first ten days of June its blog received more than 70,000 visits (25,000 in a single day) (Skytruth, 2010). Citizen journalism then appears to have played an important role in the BP oil disaster and the documenting of alternative 'lay' sources of evidence of ecological impacts.

Summary

In November 2012 BP pleaded guilty to 14 criminal charges and agreed to pay $4.5 billion in fines and other penalties. BP also agreed to pay $525 million to settle civil charges by the Securities and Exchange Commission that it misled investors about the flow rate of oil from the well, and is subject to four years of government monitoring of its safety practices and ethics (Krauss & Schwartz, 2012). At the time of writing the trial is currently ongoing.

BP's approach to the Gulf of Mexico spill is widely regarded as a public relations disaster. While they had greater information subsidies they failed to capture public trust. The company spent huge sums of money on TV advertising and bought up internet search terms but in the final analysis this was not enough to win back public support. This case illustrates the growing importance of user-engagement, opinion-influencers and interactivity when attempting to influence the popular framing of an issue. Both environmentalists and BP used celebrities and crowdsourcing but groups like Greenpeace could draw upon a much wider network of support and relationships that have been built up over time.

Environmentalists and concerned citizens effectively utilised mass self-communication to increase awareness but offline media proved to be important as well. While such strategies certainly raised visibility and challenged the official version of events the key question is how far this impacted on policy and fundamentally changed attitudes towards oil drilling. The oil company Shell is still pursuing its expansion of oil drilling into the Arctic. In a recent UK poll that asked respondents to name the top companies they would most associate with taking environmental sustainability seriously two-thirds could not name one but, of those people who could, BP came out near the top at number four (YouGov, 2013). A YouGov/*The Economist* poll conducted in 2011, one year after the spill, found that the majority of Americans surveyed (57 per cent) favoured increasing offshore drilling for oil and natural gas, with only 30 per cent in opposition (YouGov, 2011). A more recent Harris poll undertaken in 2013 suggests even higher levels of support, finding 67 per cent of voters nationwide supported offshore drilling for domestic oil and natural gas resources (Straessle, 2013). There is a danger that visibility is interpreted as influence. It remains to be seen how far such campaigns incrementally impact on policy and attitudes in the long term.

Further reading

Anderson, A. (2009) 'Communicating Chemical Risks: Beyond the Risk Society' In Eriksson, J., Gilek, M. and Ruden, C. (eds.) *Regulating Chemical Risks: Multi-disciplinary Perspectives on European and Global Challenges*. London: Springer.

Juhasz, A. (2011) *Black Tide: The Devastating Impact of the Gulf Oil Spill*. Hoboken, NJ: John Wiley.

Merry, M. K. (2014) *Framing Environmental Disaster: Environmental Advocacy and the Deep Horizon Oil Spill*. London: Routledge Chapman Hall.

Wilson, A. (1992) *The Culture of Nature: North American Landscape from Disney to the Exxon Valdez*. Cambridge, MA: Blackwell.

6
Emerging Technologies

In August 2011 a parcel bomb was sent to two nanotechnology researchers working in a university lab in Mexico City. A group named 'Individualidades tendiendo a lo salvaje', translated as Individuals towards the Savage or Wild (ITS), claimed responsibility publishing a 5,500 word attack on nanotechnology online. The bomb failed to detonate properly so fortunately there were no fatalities; however, one researcher was left with a piece of shrapnel pierced into his lung, and the other suffered burns to his legs and a burst ear drum. Had the bomb gone off as intended it contained enough explosives to collapse half the building. This was not an isolated incident. Earlier that year the group claimed responsibility for a bomb attack at a university in Mexico and it has been linked to attacks in France, Spain and Chile. And in 2010 another group, Olga Cell of the Informal Anarchist Federation International Revolutionary Front, attempted to bomb IBM's nano lab in Switzerland and claimed responsibility for the non-fatal shooting of the head of a nuclear engineering company in Genoa, Italy (Phillips, 2012). Further threats against Mexican scientists working on bio and nanotechnology were issued in 2013. In its manifesto the group claimed: 'they must pay for what they are doing to the Earth' (Beckhusen, 2013).

ITS is against industrial civilisation proclaiming 'nature is good, civilisation is evil' and they express grave concerns about synthetic biology, genetic engineering, cloning and geo-engineering as well as nanotechnologies. This distinction between the 'artificial' and the 'natural' is also evident in their outlook on digital technologies. They proclaim:

> Every day we realize that human beings are moving away more dangerously from their natural instincts, that they are immersed in

a false reality constructed by social networks and the obsessive idea of online updating in virtual spaces. We live in the digital age, the system is always in constant dynamism and not only has everyone alienated themselves through television or the vices that civilized life contracts, but also, a giant computer network has been made for the daily super-production of more automatons who serve it blindly to maintain the prevailing order.

(http://anarchistnews.org/?q=node/15216)

Ironically, their communiques can be found on the web. The internet has provided eco anarchists new opportunities to develop networks across the world and a degree of control in their self-representation. Such groups seek to promote an alternative counter-hegemonic discourse through violent direct action, designed to fundamentally challenge society's values (see Melucci, 1981). As DeLuca observes:

Progress and nature, along with the other ideographs in the discourse of industrialism, define our society for us, justify certain beliefs and actions, and signify collective commitments, such as the belief in the necessity and possibility of unlimited growth, the belief in technology as the answer to all problems (including spiritual and environmental problems), and the treatment of all nonhuman life forms as resources to be exploited (certain groups of humans get defined as nonhuman in certain circumstances).

(1999: 48)

This 'anarcho-primitivist' perspective, as it has been called, appears to be influenced by radical green thinkers such as Derrick Jensen. The group is thought to have first protested against Mexico's political and economic situation by setting off small explosives at bank machines but, with mounting social unrest and violence in the country, increasingly turned its sights on 'environmental contamination'. Nanotechnology may also be seen as a symbol of industrial development as Mexico has put a very high level of investment in this area compared with other developing countries. Underlying this extreme response to developments in science and technology is the perception that citizens have no control or say about decisions that affect the natural environment and are seen as exacerbating rather than reducing social inequality. Thus so far no risk event or controversy involving nanotechnologies has emerged on the scale of that accompanying biotechnology or climate change, although safety concerns have arisen in China and Germany (see Anderson et al., 2009; Satterfield et al., 2013).

The digital revolution then has created new ways of visualising and responding to emergent technologies (see Figure 6.1). While there may be little exposure through mainstream media, social media offer the space to project alternative views.

The rise of the internet has allowed a rapid redefinition of the significance of emerging technologies. As we shall see, it tends to offer more negative risk-oriented framings. As DeLuca notes: 'With the rhetorical tactic of image events, radical environmental groups are contesting the hegemonic discourse of industrialism and the received meanings of the ideographs progress, nature, humanity, reason, and technology' (1999:51–2).

News media attention (and silence) can play a potentially key role in influencing what we know about, and how we view, emerging technologies (Priest, 2012). Environmental organisations such as Greenpeace, FoE and Canadian Erosion, Technology and Concentration Group (ETC) are increasingly focusing their attention on the future implications of

Figure 6.1 No Nano!
Source: Wikimedia commons No Nano Grenoble P1150729.jpg

the convergence of new technologies. There are concerns that they will radically change what it means to be human and fundamentally alter the 'natural' world. This chapter discusses emerging environmental issues linked to nanotechnology, biotechnology and biodiversity, and considers the lessons that have been learnt from the reporting of previous controversies, such as genetically modified organisms. Many technologies are converging – nanotechnology, biotechnology, artificial intelligence and robotics, for example. What is the nature of debates concerning these issues and what is the potential role of the media in representing potential benefits and risks? This discussion draws upon the sociology of expectations, stigma, attention cycles and framing theory. While there are serious concerns over the environmental effects of such technologies they are seen to form an integral part of the solution to many environmental problems.

Defining nanotechnology and synthetic biology

Both nanotechnology and synthetic biology are crosscutting interdisciplinary fields and their very definition has sparked much debate. As Schummer notes in relation to nanotechnologies:

> There is no doubt that in the fields called 'nano' numerous scientific and technological breakthroughs have been made and that cutting-edge research is being done. Yet, it is difficult to specify what they have in common other than meeting standard definitions of nanoscience and nanotechnology, which are notoriously vague.
> (2005: 163)

Nanotechnologies have an incredibly diverse array of applications stemming from many different scientific disciplines resulting in everyday consumer products. There is already a range of products on the market containing engineered nanoparticles including: sunscreens, cosmetics, golf balls, tennis rackets, textiles, paints and computer hard drives. Nano-materials are already being used in food packaging, for example, nano-composites in bottles to minimise carbon dioxide leaking out. Developments are already underway to use nano-sensors in plastic packaging (that can detect when food has gone off and change colour to alert consumers) and plastic films containing silicate nanoparticles designed to give food a longer shelf life.

At its simplest, nanotechnology can be defined as the design and manipulation of matter at the atomic or molecular level. A nanometre is one billionth of a metre, which is about 100,000th of the width of a human hair. The very large surface area to volume ratios of extremely small nanoparticles gives them special properties that do not occur with larger sizes. For example, Babolat's VS Nanotube Power racket, brought out in 2002, was made of carbon nanotube-infused graphite which made it much stronger than steel yet exceptionally light. Nano materials have the potential to substantially increase the energy efficiency of electronic devices and advance 'zero-waste' technologies to support environmentally sustainable initiatives. They can also be formed into very thin films that can be used for environmental cleaning and sensing. There are already thousands of products that incorporate the use of nano materials although it is very difficult to be certain about the precise number (see Project on Emerging Technologies Inventory http://www.nanotechproject.org/inventories/consumer/). While many supporters of nanotechnology see it as heralding the next Industrial Revolution, some critics present a vision of self-replicating nanobots taking over the planet. A less extreme view is that manufactured nanoparticles may pose a real toxic threat if ingested or unleashed into the environment (Handy & Shaw, 2007; Selin, 2008).

Synthetic biology is similarly hard to define and there is no one accepted definition. According to the UK Royal Academy of Engineering: 'Synthetic biology is an emerging area of research that can broadly be described as the design and construction of novel artificial biological pathways, organisms or devices, or the redesign of existing natural biological systems' (2009b: 6). The ability to artificially synthesise DNA, and therefore create DNA parts, has been the most significant technical advance so far. In 2010 Craig Venter and his team produced the first synthetic bacterial genome and they used this to take over a cell. As with nanotechnologies synthetic biology has many environmental applications. Proponents argue that it will produce new types of environmentally friendly pesticides, generate more advanced biofuels from renewable resources, purify water through the use of advanced biosensors, and it is claimed that artificial leaf technology could help in reducing CO_2 emissions.

However, a number of issues have been raised over potential risks to human health and the environment, the dominance of large multinational corporations, and ethical questions around interfering with nature. A particularly serious concern is the potential contamination of the environment by intentional or accidental release of organisms, or their use in bio-warfare. A feature of many emerging technologies is a fundamental lack of certainty about their future implications – which in many cases simply cannot be known. There is also considerable disagreement on whether they can be considered to be novel developments or simply repackaged (see Anderson et al., 2009; Kronberger, 2012; Zhang et al., 2011). There are concerns that nanotechnology and synthetic biology have been rebranded to attract funding and scientists are divided on whether they constitute a new area of science or not (Schmidt et al., 2008). Considerable amounts of money have been invested in such technologies in the UK and elsewhere (see Figure 6.2).

Emerging technologies pose particularly difficult challenges for governance given the convergence of different disciplines and the cross-bordered nature of their reach and a comprehensive regulatory apparatus has yet to be developed. As Zhang et al. observe: a 'tangled web of global stakeholders is involved' (2011: 28). Currently synthetic biology and nanotechnologies are self-regulated by voluntary guidelines developed by industry and government. However, there are concerns that voluntary regulation is inadequate. The complexities and fragmentation of the nanotech field pose particularly tough challenges for regulation and there are fears that some nano materials could completely bypass testing and safety evaluation. Regulation cannot keep up with the sheer pace of developments. Some NGOs (such as the ETC group in Canada, which switched their major focus from agri-biotechnology to nanotechnology) have called for a moratorium, and the insurance industry has called for greater regulatory oversight. Similarly with respect to synthetic biology, environmental groups such as FoE and ETC have raised a number of serious concerns over regulation (FoE, 2012).

Public awareness of emerging technologies

In the main there has been very little resistance to nanotechnologies worldwide. Public awareness of nanotechnologies remains relatively low and attitudes are generally positive, although there is less acceptance of the idea of nanoparticles in food and food-contact materials (Hart Research Associates, 2013; Harris, 2012; Satterfield et al., 2009; Siegrist et al., 2007; Vandermoere et al., 2011; Zimmer et al., 2008). Research suggests that when assessing new technologies people tend to draw upon value systems and predispositions influenced by earlier controversies (see Figure 6.3).

Public trust is vital with such a sensitive topic as food, particularly in Europe where the handling of the crisis over genetically modified (GM) crops appears to have deeply impacted upon attitudes towards the use of nanotechnology in foods (although there appears to be much greater fear attached to its application directly to food). In contrast, evidence suggests that US citizens have tended to hold more positive perceptions

134

HM Government   Eight Great Technologies

🧪 Synthetic Biology

Harnessing the power of biology to fuel and heal us

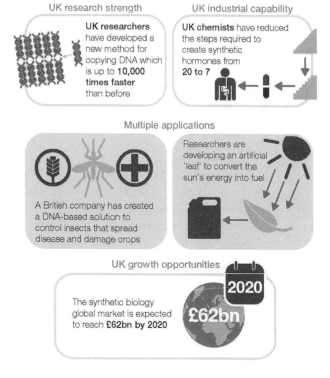

UK research strength

UK researchers have developed a new method for copying DNA which is up to **10,000 times faster** than before

UK industrial capability

UK chemists have reduced the steps required to create synthetic hormones from **20 to 7**

Multiple applications

A British company has created a DNA-based solution to control insects that spread disease and damage crops

Researchers are developing an artificial 'leaf' to convert the sun's energy into fuel

UK growth opportunities

The synthetic biology global market is expected to reach **£62bn by 2020**

£62bn

2020

Accelerating the commercialisation of technologies   #8Great
www.gov.uk/bis/industrial-strategy

Figure 6.2   Accelerating the commercialisation of technologies
Source: Flickr,   http://www.flickr.com/photos/bisgovuk/10210847256/sizes/c/in/photolist-
gyig9C-fmFbZG-8qaiNE-fmzMcA-fjRKhY-8RH4qX-7M5zua-dR6coi-g6NGRm-jpMSTV-
8WK7Vr-8WNawj-8WK7DZ-8WK7KP-8WK7Y6-8WK84k-8WNahU-8WK81z-8W/

Figure 6.3   Occupy Monsanto protest
Source: Flickr,   http://www.flickr.com/photos/69387260@N02/6846151608/in/photolist-
bqYjmG-etPhGT-etSBLo-etSAyC-etPmN2-etSzjo-etSuxf-etPeEz-etPkGF-8bjb2U-eLsfV8-
etSrE9-etPo7z-euDgWE-euCVm9-eLsfMp-etyCbs-etuNgB-gCRSJz-etu2yR-k7EyK8-i2b6Wc-
ec1Udb-etx9f9-gYjLag-ex5Xcg-bNqt82-bzRbnt

of 'nano-food' and nano packaging than their European counterparts (Lyndhurst, 2009).

Despite various attempts in the US to engage the public, levels of knowledge have stayed broadly the same over recent years but this has not prevented people from developing opinions on nanotechnologies (see Cacciatore et al., 2011). The 'Europeans and Biotechnology: 2010' report found that 45 per cent of EU citizens claim they have heard of nanotechnology. Six out of ten respondents who expressed an opinion said they supported nanotechnology applications – with the highest level of support (over 70 per cent) in Poland, Cyprus, the Czech Republic, Finland and Iceland, and the lowest level of support (less than 50 per cent) in Greece, Austria and Turkey (Gaskell et al., 2010). Other research undertaken in Japan suggests that Japanese people are the most optimistic of all about nanotechnologies, with around 80 per cent viewing developments as beneficial (Fujita et al., 2006; Kishimoto et al., 2010).

Similarly public awareness about synthetic biology in the UK and US still appears to be very low. Publics appear to feel kept in the dark about these developments (BBSRC, 2011). Studies undertaken in 2008 and 2009 suggested that two-thirds of people had not heard about it at all (see Hart Research Associates, 2008; Pauwels, 2009; RAE 2009a, 2009b). When the Royal Academy of Engineering asked respondents what words come to mind in relation to synthetic biology almost half (49 per cent) could not say or did not know. Of those who could associate particular words the most frequently mentioned were 'man-made', 'synthetic', 'biology', 'genetic' and 'cloning' (see Figure 6.4).

A European Commission survey undertaken in February 2010 found that 83 per cent of Europeans reported that they had not heard the term (Gaskell et al., 2010). Those who said they had heard about synthetic biology were much more likely to approve of it, provided there is strict

Figure 6.4 Synthetic biology associations word cloud
Source: Royal Academy of Engineering (2009a) *Synthetic Biology: Public Dialogue on Synthetic Biology*, p. 19

regulation, while those who said that they were not aware were more likely to respond that they did not know whether it should be permitted or that they would not support it under any circumstances. A survey of attitudes towards science that was conducted in the UK in 2011 asked people about the risks and benefits attached to synthetic biology, to which 35 per cent responded that they did not know enough to answer the question and 87 per cent responded that they felt uninformed about the topic (IPSOS-MORI/BIS, 2011). Around a third (33 per cent) perceived the benefits as outweighing the risks and 12 per cent saw the risks as outweighing the benefits. With regard to both nanotechnology and synthetic biology the findings from UK survey research suggests that women tend to feel less informed than men (Sciencewise, 2013).

A more recent survey of US adults found that less than a quarter of respondents (23 per cent) claimed that they have heard a lot or something about synthetic biology, while slightly more said the same about nanotechnology (32 per cent). They most commonly associated it with 'unnatural', 'artificial', 'man-made' or that it is to do with reproducing life (Hart Research Associates, 2013). A clear majority of respondents supported synthetic biology in general; 61 per cent thought that it should be allowed to develop rather than be banned until more is known about the benefits and risks. However, attitudes varied according to specific applications such as use in new food additives or as a crop enhancing fertiliser. In general most of the respondents thought that the risks and benefits of synthetic biology are about the same (40 per cent), although 27 per cent stated that they did not know. Once all respondents were provided with more information about synthetic biology they were more likely to see the risks outweighing the benefits, rising from 15 per cent (before being informed of potential benefits and risks) to 33 per cent afterwards. The numbers believing benefits would outweigh the risks only rose by 6 per cent once further information was provided. Most still thought that the risks and benefits would be about equal (38 per cent). University scientists and researchers were by far the most trusted people to maximise the benefits and minimise the risks. The highest level of concern was expressed in relation to the potential to develop biological weapons and that it is morally unacceptable to create artificial life, and some saw potential negative health impacts as an issue. Relatively few respondents saw environmental impacts as a concern (12 per cent). Women, the less highly educated and those affiliated with a religious organisation were more likely to express concern. Views on regulation were divided as to whether the federal government should be involved or whether there should be voluntary guidelines. Likewise studies of public attitudes to nanotechnologies suggest that females and those who are more religious express greater concerns (e.g. Cacciatore et al., 2011).

Despite being low, levels of public awareness appear to be rising in the USA; indeed, between 2008 and 2010 the number of people reporting awareness of synthetic biology increased three fold (Pauwels, 2013). Most research to date has been quantitative in nature, but there is some qualitative work that is emerging. Eleonare Pauwels, for example, has undertaken focus group research in the US in order to delve deeper into public attitudes (see Pauwels, 2013). This suggests there is considerable ambivalence. Upon being given more information about the potential risks and benefits of synthetic biology focus group participants were more likely to view it in more negative terms.

Surveys suggest that there are significant cultural differences in support for emerging technologies. However, we need to approach survey findings with a measure of caution since they vary in terms of question wording and how different terms are translated into different languages (Priest, 2012). A number of different explanations could be offered for apparent differences. The use of surveys to assess levels of trust in nanotechnologies or synthetic biology is problematic in a number of ways. Nanotechnologies, for example, are largely invisible and there are relatively few everyday analogies that can be drawn upon to make the field meaningful to people (Pidgeon et al., 2011). Surveys may skew people's attitudes by forcing them into neat compartments and obscuring nuances of meaning; particularly given that there is much uncertainty and ambiguity over the concept of 'nanotechnology' itself (Anderson & Petersen, 2012; Åm, 2011; Kearnes et al., 2006). As Pidgeon et al. observe: 'Surveys cannot easily uncover the ways that people will interpret and understand the complexities of nanotechnologies (or any other topic about which they know very little) when asked to deliberate about it in more depth' (2009: 95).

Despite having little information about nanotechnologies, people form opinions on the basis of prior experience and previous controversies involving emerging technologies. As Cacciatore et al. argue:

Given the role that nanotechnology plays as an enabling technology, the mental associations one has with the technology can play a potentially powerful role in shaping how risk and benefit information is utilized when forming opinions. These associations may be to the medical field, the military, or possibly, to tiny self-replicating robots. While some of these associations may point to the benefits of the technology, others may make salient a wide variety of risk factors and this can have huge implications for how individuals form opinions about the technology.

(2011: 388)

More in-depth qualitative research using a participatory methodology suggests that when people weigh up the benefits and risks of technology, they tend to draw upon a generic model of 'new technology' that tends to be interpreted as beneficial to society (Pidgeon et al., 2011). This would seem to reflect the dominance of utopian representations within much news media coverage of emerging technologies (Anderson et al., 2005). Nevertheless, there is still the potential for conflict and resistance once the issues become contested if it moves into a more overtly political arena. Indeed, the types of frames that follow an issue appear to be significantly influenced by its stage in the issue-attention cycle (Downs, 1972; Nisbet & Huge, 2006). For example, frames focusing on new scientific discovery dominated early news media coverage of biotechnology and stem cell research and little consideration was given to social implications (Nisbet & Huge, 2006). In the case of biotechnology it was only in the late 1990s that it started to attract more critical coverage. As Nisbet and Huge (2006) observe, if issues become more politicised and media attention picks up then frames highlighting moral and ethical aspects are more likely to appear as claims-makers seek to dislodge the issue away from the administrative policy arena to a more overtly political arena. The shift in framing in technical terms to one of drama, in turn, means it is more likely to be reported on by political or generalist reporters rather than specialist science correspondents. Indeed, Nisbet and Huge (2006) found that conflict and political strategy frames were most common when news media coverage peaked. This suggests that we can identify a general pattern whereby traditional news media coverage is generally positive in the early stages and becomes more conflict-driven over time, if the right ingredients are present to make underlying issues more dramatic.

The potential for stigma

Stigma, which occurs when the potential of a technology becomes tarnished by discourses of risk, is very difficult to overcome once it takes hold (Wilkinson et al., 2007). Gregory et al. define it in this way:

Stigma goes beyond conceptions of hazard. It refers to something that is to be shunned or avoided not just because it is dangerous but because it overturns or destroys a positive condition; what was or should be something good is now marked as blemished or tainted... Technological stigmatization is a powerful component of public opposition to many proposed new technologies, products and facilities.

(2001: 3)

The scale of nanotechnology means that risks are 'non-visible' to the human eye. Nanoparticles are viewed as un-bounded and thus may involve indiscriminate exposure to the natural environment. An Editorial in *Nature* warned:

New results from the field of nanoscale science are paraded into the media spotlight almost daily, each allegedly capable of curing disease, cleansing the environment, or otherwise extending peace and prosperity. Such hype ensures that nobody will believe the message that nanotech is old news. And if the technology is capable of such wonders, people may muse, the chances are that it can probably do bad things as well.

(2003: 237)

The Synthetic Biology Dialogue found that stakeholders drew analogies between synthetic biology and the GM debate (BBSRC, 2011). From the UK focus groups they found that there was conditional support for synthetic biology; there was considerable enthusiasm for the potential of synthetic biology, but fears about issues of control and misuse, who benefits and how regulation would be enforced, as well as impacts on health and the environment (see BBSRC, 2011). In the UK context there are particular concerns about a potential re-run of the crisis over GM food and crops.

Synthetic biology represents just one illustration of how rapid advances in the life sciences are opening up a host of potentially dramatic new applications in medicine and healthcare, agriculture, industrial chemistry and energy production, among other fields. These developments also introduce possible new risks that are necessarily speculative and hard to assess. Yet there is often pressure for decision-making about risk governance and regulation to begin many years before actual products appear on the market. Errors of judgement at these early stages can have a major impact on the trajectory of a new technology, as well as on the effectiveness and efficiency of risk governance itself.

(IRGC, 2010: 7)

Such prior controversies demonstrate how public perception can significantly impact on corporate reputation and profit, and it is clear that risk analysis must be conducted in an inclusive and transparent way. While news coverage is currently low-key, anxiety over the possible toxic effects of nanoparticles (invisible to the human eye) or synthetic biology has the potential to trigger major alarm. Yet there are many uncertainties and past historical experiences with new technologies are not always a good guide to the future (Selin, 2008; Torgersen & Hampel, 2012).

In the UK the primary source for the potential stigmatisation of nanotechnology thus far was the coverage of HRH Prince Charles's purported comments in 2004 on the so-called grey goo scenario (see Anderson et al., 2005). However, the Prince's comments, which appeared in the run up to the release of the Royal Society study, had the effect of ensuring that governmental precautions were not widely identified as a trigger for the social amplification of risk discourses in the media (Kasperson et al., 2001). Instead when the associated 'grey goo' controversy proved to be fleeting, media attention quickly receded. Indeed, as Arnall and Parr observe: 'not only do such concerns ['grey goo'] deflect attention from short and medium-term issues that demand more immediate attention, but they are also easy to dismiss as 'fanciful', detracting from real issues that are emerging' (2005: 25).

The media, particularly television, have played an important role given that they constitute the main source of information for people on science after formal education (Bubela et al., 2012; IPSOS-MORI, 2011). As Bubela et al. observe: 'Although media effects on public opinion are generally overstated, the media can focus public attention on or away from specific issues (agenda-setting) and can frame issues to the benefit or detriment of specific stakeholders' (2012: 133). As discussed in Chapter 3, frames are influenced both by journalistic news values and by news sources that pursue their own particular agendas, and inevitably offer a partial and selective view of reality.

News sources and the struggle for power

Recognising the significance of media coverage for shaping policy agendas and public attitudes, news sources (e.g. scientists, NGOs, government, industry) compete to get their voices heard and influence the reporting in a way that best reflects their interests. The media play a potentially crucial role in framing newly emerging issues, mainly by helping to establish the initial parameters of debate, identifying particular news sources as pertinent and credible, and by providing topic-defining reference points (Priest, 1994). As Nisbet et al. observe:

> Not only do the media influence the attention of competing political actors and the public but the media also powerfully shape how policy issues related to science and technology controversy are defined, symbolized, and ultimately resolved... Once an issue is framed or characterized early on in a debate by the media, it can be very difficult for policymakers or other interests to shift the image of the issue to another perspective.
>
> (2003: 38–42)

As pointed out in previous chapters certain news sources are likely to have more success in framing an issue than others. Prior studies of biotechnology controversies found that framing tended to have an overwhelmingly positive spin and scientists, industry and government officials tended to be the most prominent sources in news media coverage (Nisbet & Lewenstein, 2002). Gutteling et al.'s (2002) study of biotechnology coverage in opinion-leading newspapers in 12 European countries over the period 1973 to 1996 revealed that 'scientists' and 'industry' constituted more than half the references to actors in most countries during this time span, while 'politicians' and 'NGOs' were much less prominent. However, there are differences in the relative prominence of the various actor groups across different European countries. Scientists are frequently quoted or cited in news stories on medical genetics, which may help explain the generally positive portrayal of issues found in a number of studies (Conrad, 2001; Gutteling, et al., 2002; Kitzinger, et al., 2003; Petersen, 2001). Bubela and Caulfield (2004) found that in 95 per cent (591 of 627) of newspaper articles on gene discoveries published in Canada, US, UK and Australia the main source of information cited was the scientific paper or its authors. The authors discovered that opinions outside the research community were sought in only 8 per cent of newspaper articles and these opinions were the primary voice in only 2 per cent of articles (2004: 1401). However, as Ten Eyck and Williment (2003) argue sources do not occupy a homogenous, hegemonic position and trajectories can alter. Crises or dramatic events (e.g. the announcement of the birth of the cloned sheep Dolly) may lead to a shift in source influence, allowing otherwise perhaps politically marginal groups (such as community activist groups) to gain prominence in framing debates (Nisbet et al., 2003). Gutteling et al.'s (2002) European study, noted above, revealed differences between countries in the reliance placed upon various actors and stakeholders over time. The findings were found to parallel those from an earlier study in the US (Priest, 1994) and a study incorporating US and European data (Kohring et al., 1999).

Many stories are source generated – some estimates put it as many as half or more newspaper stories – so scientists are able to strategically package news items for journalists (Nisbet & Lewenstein, 2002: 362). Journalists increasingly lack the time, means or expertise to seek independent verification of facts and are often reliant on pre-packaged information over which they have little control (Lewis et al., 2008; Manning, 2001). Many scientists are aware that the continuity of their research funding is heavily reliant on public support that can be influenced by media coverage of science research. Hence, many are keen to use the media to promote a positive image of research and of scientists (Burke, 2003; Nelkin, 1985, 1994). By controlling the timing of news releases, and by choosing particular language and metaphors, scientists may seek to shape public discourse and influence the direction of policy (e.g. Nelkin, 1985). With controversial issues scientists have an opportunity to play a major role in influencing public understandings of science, not least through their ability to select voices (Dunwoody, 1999: 69). They may use conflicting opinion and uncertainty to their advantage, as a rhetorical tool. That is, they may use the occasion to highlight the validity of their own work and the uncertainties of others with whom they disagree (Dunwoody, 1999: 73).

One analytical framework for understanding processes of societal conflict over emerging technologies is the 'Gate Resonance Model' developed by Torgersen and Hampel (see Torgersen & Hampel, 2012). The model attempts to conceptualise the interaction between public opinion, interest representing organisations, the media and the political system. Like Hilgartner and Bosk's 'Public Arenas Model', this analytical framework recognises that there are several different arenas over which conflict is played out and that media constitute multiple platforms with their own particular carrying capacities (Hilgartner & Bosk, 1988). The Gate Resonance Model emphasises the role of the regulatory system as gatekeeper:

> They determine the success of interest representation by selecting the issues they deal with and restrict the frame of the regulatory debate. They also select interest groups regularly involved in political negotiation processes (so-called established interests), while others (organized interests) are excluded. In addition to selecting interest groups,

institutions select problems to deal with and types of arguments that are either processed or rejected . . . Any interest-representing organization not integrated into the normal mode of policy-making can either transform their views in terms of fitting into the Gate or try to modify it, for example via mobilisation strategies. Resonance with the media and the public is an important prerequisite for mobilisation, but the number of issues taken up is limited and subject to attention cycles and competition.

(Torgersen & Hampel, 2012: 136)

Instead of being static it recognises that the influence of actors changes over time. It also recognises the existence of different 'publics'; that is, it makes the distinction between specific groups who hold particular interests in common (e.g. farmers, students) which may overlap with, but is distinct from, those who share similar worldviews (e.g. political ideologies or religious beliefs). Torgersen and Hampel argue that when official policy and regulatory institutions take on board the need to discuss ethical, legal and social issues it offers less of an opportunity for organised interest groups to mount campaigns around this and mobilise publics. This, they suggest, is one reason why synthetic biology has not generated much interest thus far, in contrast to the GM conflict where NGOs gained more of a foothold in influencing policy. Much rests then on the ability of non-established interests or 'challengers' to circumvent the gate function and gain access to the regulatory system. For this to happen there has to be some sort of trigger that mobilises widespread support from unlikely alliances of groups representing opposing interests. This usefully draws attention to inequalities of power among news sources, and the complexities and contingencies that differentially impact upon the trajectory of debates concerning particular types of emerging technologies. One important difference between the conflict over GM crops and that of nanotechnologies or synthetic biology is that attempts to involve citizens and stakeholders have occurred much earlier on and it is to this that we turn to discuss in the next section.

Public engagement and the deficit model

As mentioned above, a number of concerns have been raised over the possibility of a repeat of the GM crisis with nanotechnologies and synthetic biology should a major trigger event occur (Anderson et al., 2009; Bhattachary et al., 2010; Kronberger, 2012). In particular some have cautioned that the hype associated with many emerging technologies may be counter-productive and lead to 'disproportionate social, ethical and regulatory responses' (Bubela et al., 2012: 132). From the outset proponents of nanotechnology research and development have been concerned about managing public responses to the field so an upstream public engagement approach has been developed in many European countries (IRGC, 2010; Tait, 2009; Torgersen & Schmidt, 2013) – see Figure 6.5. The concept of 'upstream public engagement' emphasises the need to involve publics and stakeholders in the very early stages so that their views can influence the path technological developments take. This reflects concerns that with the GM crisis public dialogue was undertaken much too late in the process, so it afforded no opportunity to influence how the products and technologies were developed in the first place.

The concept of upstream public engagement has generated considerable debate. It is often used in an over-simplified way and is not intended to mean merely discussion taking place in advance but technological development being informed by public opinions (Rogers-Hayden & Pidgeon, 2007). Some have questioned whether it really has genuinely opened up debate to a variety of views from different publics and stakeholders. Tait, for example, argues: 'upstream engagement seems likely merely to substitute one set of dominant opinions for another set that is no more universal, and if anything, is less based on scientific evidence than the previous one' (Tait, 2009: S21).

Certainly the notion of upstream public engagement employed by some moves us little forward from the traditional linear deficit model of the public understanding of science. The question of how technologies

'Dialogue and deliberation, that includes the publics and related interest groups, relevant science communities and policy makers, about potentially disruptive/controversial technologies at an early stage of the research and development process and in advance of significant applications, or widespread public knowledge, in a way that has the potential to influence the technology trajectories.' (Rogers-Hayden & Pidgeon, 2008: 1011).

Examples in the UK include:

- Nanodialogues
- NanoJury UK
- Small talk
- Nanotechnology Engagement Group

Figure 6.5   Upstream public engagement

are represented and how this may influence public responses and ulti-
mately public trust has become a crucial issue in the contemporary
governance of new and emerging technologies (Anderson et al., 2012;
Anderson & Petersen, 2012). The deficit model, the idea that all we
need to do is provide publics with more knowledge and this will generate more support, has been widely discredited (see Anderson, et al.,
2009). There are a number of recurrent, questionable assumptions about
the science-society relationship that have guided science communication efforts in this area. One such assumption is that greater or 'better'
or more 'accurate' information about the science will resolve communication difficulties and facilitate 'trust'. Underlying many of the
policies in this context is trust that innovations will evolve as and
when they are envisaged, that benefits will be distributed widely and
that risks are governable. Given the general lack of familiarity with
nanotechnologies, which for many may be viewed as a 'non-issue',
we cannot simply apply the same frameworks of trust deployed to
understand previous biotechnology controversies. Indeed, there is little
appreciation of how scientists' own framings of emerging technologies have the potential to impact on subsequent policy and public
responses (Anderson & Petersen, 2012). As Priest points out, a beneficial 'technology template' tends to dominate people's interpretation
of nanotechnologies and the discourses associated with them, whereas
with biotechnology views tended to be sharply polarised from the very
start (Priest, 2012).

Providing people with more information does not necessarily lead
them to be more likely to approve of new technologies. In fact, evidence
suggests that in some cases it can result in them becoming *less accepting*
(see Pauwels, 2013). And in countries where scientific literacy is generally high there can be considerable opposition to some controversial
technologies such as GM food (Bubela et al., 2012). An alternative view
is to see the deficit not so much as lying with public knowledge but in
terms of a lack of trust in experts and institutions (Anderson & Petersen,
2012; Kronberger et al., 2012). While they point to different causes of
the uneasy relationship between science and society these perspectives
share an underlying hope that public engagement will lead to greater
support.

People do not approach new technologies, even if they have no prior
knowledge of them, as 'empty vessels' waiting to be filled (Anderson &
Petersen, 2012; Kronberger et al., 2012). Studies demonstrate that we
anchor new technologies in our prior knowledge of other emerging
technologies (Kronberger et al., 2012; Torgensen & Hampel, 2012). Risk
perception is influenced by past associations; people's assessment of the

potential negative consequences of nanomaterials was greater where a 'multinational corporation' scenario was used rather than a small enterprise (Schütz & Wiedemann, 2008).

In order to process new information we use 'schemata of interpretation' to help grasp the complexities and ambiguities of everyday life and particular discursive news frames resonate more closely with these themes than others. As Trond Grønli Åm, observes:

> if trust is to maintain its important role in evaluating emerging technologies the approach has to be widened and initially focus not on people's *motivations* for trust, but rather the object of trust itself, as to predicting how and where distrust might appear, how the object is established as an object of trust, and how it is established in relation with the public.
>
> (2011: 15, emphasis original)

The Cultural Cognition Project suggests that people who have an individualist/hierarchical outlook tend to view nanotechnologies in a more positive light, whereas those who have an egalitarian outlook tend to view it as more risky (Kahan et al., 2008; Kahan et al., 2009). However, a very different picture emerges for synthetic biology. Braman et al. (2009) conducted an online survey of a nationally representative sample of 1,500 people in the US. They found that around a half saw the benefits as outweighing the risks although a large number of people only agreed slightly with this statement, suggesting that there was considerable ambivalence. However, the Cultural Cognition Project found that those people most concerned about synthetic biology risks tended to have an outlook that was politically conservative, very religious and hierarchical. A large majority of the respondents (82 per cent) claimed to know 'little' or 'nothing' about synthetic biology but the amount of reported knowledge about it appeared to have little effect on their attitudes. This is in contrast with public attitudes towards nanotechnologies where familiarity does appear to be linked with being more likely to view the benefits as outweighing the risks (Kahan et al., 2008).

This same group displaying this cultural worldview tend to be the most sceptical about environmental issues such as climate change and nuclear power, and more likely to view nanotechnology in beneficial terms. Why should this be the case? According to Braman et al. (2009) it can be explained by the perception that synthetic biology involves tampering with nature or 'playing with God'. People respond to emerging technologies, knowing little of what they involve, with intuitive gut responses that draw upon their underlying views about authority, science and morality. Given that synthetic biology seems to trigger a different cluster of cultural meanings they conclude that synthetic biology may have particular potential for future social controversy: 'a distinctive form of cultural conflict over the risks of synthetic biology is indeed a realistic possibility' (2009: 7), although they acknowledge that further research is necessary. In particular it is not clear how far these findings are peculiar to US publics where political/religious opinions are particularly strongly divided (see Anderson, 2014).

However, we know from examining previous controversies over emerging technologies that attitudes tend to change over time. For some issues this involves a dramatic shift, while for others the issue never reaches the stage where it attracts large-scale political debate and social conflict.

Cycles of hype and hope: News media framing of emerging technologies

Publics' reactions to technologies are shaped not only by the substantive content of information, but the means by which it is packaged. The role of the media in framing nanotechnology and, more recently, synthetic biology, has been the subject of considerable debate. There have been many more studies examining press coverage of nanotechnologies than synthetic biology. By far the most studies to date examining nanotechnology coverage have focused on the US print media (e.g. Dudo et al., 2011; Friedman & Egolf, 2011; Gorss & Lewenstein, 2005; Lively et al., 2012; Weaver et al., 2009). However, over recent years there have been a growing number of studies examining European coverage, again mainly focusing on the print media. These include: Austria and Switzerland (Metag & Marcinkowski, 2014); Italy (Arnaldi, 2008); Germany (Donk et al., 2012; Gschmeidler & Seiringer, 2012); Denmark (Kjærgaard, 2010); the Netherlands (Te Kulve, 2006); Sweden (Boholm, 2013; Lemańczyk, 2013); Norway (Kjølberg, 2009); Poland (Lemańczyk, 2012); Slovenia (Groboljsek & Mali, 2012); and the UK (Anderson et al., 2005; Anderson et al., 2009; Ebeling, 2008). Further afield researchers have also focused on coverage in India (Anand & Deepa, 2013; Kanerva, 2009) as well as Hong Kong, South Africa and Kenya (Kanerva, 2009).

The above studies suggest that nanotechnologies have generally been framed in an overwhelmingly positive light emphasising the benefits rather than the risks; although more recently there appears to have been a slight shift towards greater coverage of risk and regulation, at least in the US national press (e.g. Dudo et al., 2011; Weaver et al., 2009). Over time there has been an increase in coverage, despite low public awareness of the issues, but levels have dropped off since their peak in 2006 (Cacciatore et al., 2012; Lively et al., 2012; Veltri, 2012). News interest has generally been low-key and in order for the topic to gain coverage it generally needed a hook focused around an event. For example, Friedman and Egolf (2011) found that 68 per cent of the US and 58 per cent of the UK articles they sampled were based on an event, the release of a report or statements by well-known figures.

Reports in the national press have been particularly linked to scientific advancements, especially in medicine. Given that press coverage is increasingly dependent upon PR agencies and corporate press releases this should not be a great surprise (Lewis et al., 2008). The progress frame presents nanotechnology as the inevitable outcome of technological developments – it 'just happens' – and issues over regulation and risk are downplayed (Lively et al., 2012). The progress frame appears to have been particularly dominant, especially in US news media coverage (see Dudo et al., 2011; Lively et al., 2012; Weaver et al., 2009).

Alternative frames identified by Lively et al. (2012) include the conflict frame, generic risk and, finally, regulation. For Donk et al. (2012) the main frame identified in German national press coverage was 'Research and development'. This was the most prominent frame accounting for almost half the articles and it was dominant in virtually all the years covered by the study (2000–2008). However, there are differences in framing that can be identified between different newspapers with some emphasising the economic frame, for example, much more than others. For example, in our study of nanotechnology coverage in national daily newspapers over the period 2003–2004 we found that the *Financial Times* tended to use a 'business frame' whereas *The Guardian* tended to adopt a 'social implications' or 'scientific discovery' frame (see Anderson et al., 2009).

Nanotechnology coverage focused slightly more on risks from the outset compared with biotechnology. A longitudinal study spanning ten years by Friedman and Egolf (2011) found that health risks received more prominence than environmental risks. In the UK there was more coverage of societal risks and safety issues perhaps reflecting previous experiences with bovine spongiform encephalopathy (BSE), and GM food and crops. However, our study suggests that a scientific discovery frame still predominated. We undertook a supplementary content analysis of the reporting of nanoparticle risks in the UK national press over the period 2003–2006 and the findings suggest that nanoparticle toxicology received very little attention (see Wilkinson et al., 2007).

Prior studies of emerging technologies highlight the importance of this formative period of social issue construction (Nisbet et al., 2003). Once the news media frame an issue early on in a debate, it can be difficult for policymakers to shift the perception of the issue (Anderson et al., 2009; Hall et al., 2013). In this small element of the total press coverage in our wider study just under half of the articles (15 in total) were focused on the potential toxicology of nanoparticles. Interestingly all of these articles appeared in the opinion-leading newspapers rather than the popular press. Specifically, seven stories were published in *The Guardian*, four in *The Times*, two in the *Financial Times* and one in *The Daily Telegraph*. Over this time period *The Observer* was the only Sunday newspaper to feature an article that discussed the risks of nanoparticles. Although nine of the sampled newspapers featured articles that discussed the potential toxicology of nanoparticles the coverage made up only a small proportion of the articles that appeared. Only sporadic attention was devoted to this topic by those newspapers that did cover it over the period. This supports the findings of other studies, which suggest that coverage has tended to focus upon general risks rather than more specific potential dangers, especially in relation to environmental impacts (see Friedman & Egolf, 2005).

Fewer studies have been undertaken on how synthetic biology has been represented by news media, but in recent years a growing number of researchers have begun to examine this. Such studies suggest that there has been a significant increase in the number of newspaper articles in the 'quality' press on synthetic biology in the US and Europe in recent years (Pauwels & Ifrim, 2008; Cserer & Seiringer, 2009; Gschmeidler & Seiringer, 2012). A content analysis undertaken by Synth-Ethics, an EU funded 7th Framework Programme project, found that between 2005 and 2008 the number of articles increased nearly seven-fold (see Synth-Ethics, 2010). Germany appears to have begun the trend in 2005 although by 2008 the number of articles dropped by 50 per cent (Synth-Ethics, 2010). As with nanotechnology, some significant differences can be observed across different cultural contexts in different countries. For example, biosecurity issues appear to feature much more strongly in press discourse on synthetic biology in the US than in Europe (see Pauwels & Ifrim, 2008). European newspapers were found to give more attention to potential environmental benefits, while in the US the press focussed more on advances in healthcare (see Pauwels & Ifrim, 2008). Slightly more attention was given to ethical issues by the European newspapers compared with the US. In China there is relatively little public debate about synthetic biology and news media coverage has tended to be positive (see Zhang et al., 2011).

Similar techno-futurist discourses are being promoted by actors involved in synthetic biology as with those related to nanotechnology. For example, Steven Chu, the Nobel-prize-winning physicist and US Secretary of Energy, and Phillip Ball, a science journalist working for *Nature* (Synth-Ethics, 2010), talking about nano-bio convergence. This is a field where the expectations, generated largely by the science and policy communities, have tended to run ahead of the state of the science. This reflects, at least in part, pressures from funders to demonstrate that research has tangible economic returns (Jones, 2008). The heightened expectations themselves are nothing new since they are similar to those that accompanied the rise of biotechnology (Berube, 2006; Bubela & Caulfield, 2004). Indeed, evidence suggests that early press coverage of nanotechnologies featured more negative frames, at least in the UK, compared with the first 20 years coverage of biotechnology (Anderson et al., 2005; Gorss & Lewenstein, 2005).

The hype also helps to legitimise the field and mobilise resources and actors, especially in the early stages of technological development. As Selin observes:

> expectations serve a very real palatable role in the development of nanotechnology…that is, the future is a rhetorical and symbolic place to work out 'what is nanotechnology?', but also serves a productive role that underlies everyday decision-making, alliance building and resource allocation.
>
> (2007: 214–15)

The most prominent metaphors in news media coverage of synthetic biology are to do with engineering technology/construction, creativity and design 'playfulness', and religion. Frequent references are made to engineering, building, bricks and machines (Gschmeidler & Seiringer, 2012). Two dominant frames for synthetic biology that have been identified are: (i) scientists are seen as 'playing God' and (ii) the (re)design of nature and the merging of biology with engineering (Schmidt et al. 2008, Dabrock, 2009). In the main, however, coverage has tended to be very positive. Public attitudes in the early stages are likely to be shaped by previous experiences with agricultural biotechnology (Gschmeidler & Seiringer, 2011), and stem cell research and cloning (Caulfield, 2005; Petersen, 2001). The news media are already using such associations with previous gene technologies (Kronberger et al., 2009; Gschmeidler & Seiringer, 2012) and this is likely to be a significant influence on emerging opinion (Bubela et al., 2012).

Emerging technologies online

An obvious limitation with such studies is that they only tell us about one particular media platform and indeed many have focused purely on the elite press, although a limited number have included mass circulation titles, and science journals and trade publications. Still, there has been a tendency to treat the media as undifferentiated. It has been posited, for example, that: 'the winning frames are the ones that are consistent with business interests and/or celebrate scientific advancement' (Fitzgerald & Rubin, 2010: 391, original emphasis). An important question arises as to whether this tendency varies across different media platforms.

Far fewer studies have examined nanotechnology coverage online compared with mainstream media, yet growing numbers of people are going online to find information about science (see Horrigan, 2006). At the same time, science and technology reporting in the US has experienced considerable decline over recent years and there have been cuts in the number of environment correspondents in the UK (AAAS, 2010; Mooney, 2008; PEW, 2010; Ward & Hicks, 2013). We increasingly 'graze' on media, digesting small chunks of news from multiple media platforms (Cacciatore et al., 2012) and journalists utilise blogs more and more as a source of story ideas on science and technology issues (Brumfiel, 2009; Usher, 2010). From a risk communication point of view Boström and Lofstedt argue that:

> As technologies continue to evolve at an increasing rate, regulators and risk communicators need to better understand and employ new media and modes of communicating, including peer-to-peer communications, in order to listen better, engage public imagination, and communicate risk more effectively with diverse stakeholders.
>
> (2010: 1657)

Cacciatore et al. (2012) undertook a study of nanotechnology coverage in 21 US newspapers compared with online content via Google News and Google Blogs between 1 January 2004 and 31 December 2009. They found that while print media coverage had started to tail off Google Blog coverage consistently increased over the sample period. Moreover, there was significantly more coverage devoted to environmental and risk related content online compared with the traditional newspaper press. Indeed, risk related content in Google search results grew by 80 per cent from April to June 2009 and thereafter stayed constant. These findings are supported by a study undertaken by Runge et al. (2013) who used computational linguistic software to examine a census of all English language tweets on nanotechnology expressing opinions posted on Twitter between 1 September 2010 and 31 August 2011. Their findings suggest that 55 per cent of tweets expressed certainty and 45 per cent expressed uncertainty. The majority of tweets expressed pessimistic views (41 per cent), against 27 per cent of tweets that were optimistic and 32 per cent that were neutral. However, Veltri's (2012) study of English language tweets on nanotechnology over a period of 60 days found that most were announcing events or medical applications; only about 15 per cent of the sample was to do with concerns and opposition towards nanotechnology. These more negative tweets mainly focussed on the potential toxicity of nano materials and military applications, but they tended to emphasise uncertainty rather than direct hostility.

Further evidence that references to the risks of nanotechnologies may be increasing online is provided by Böl et al. (2010). They undertook a qualitative analysis of German language discussions in online forums and blogs about the consumer applications of nanotechnology since the beginning of the decade, comprising approximately 500 individual posts. The study found there was a generally high level of acceptance of nanotechnology products by these active German internet posters, but references to risk and negative perceptions were judged to be increasing over time. The authors highlight a potential for conflict to emerge around food or cosmetics given the kinds of imagery being used to represent nanoproducts.

There is some evidence then that a preoccupation with risk (or at least a greater emphasis on uncertainty) may be found in online sources. The case involving the deaths of two nanotechnology printing factory workers in China in 2009 that suffered severe lung damage, allegedly caused by inhaling silver nanoparticles, received much greater and more sustained coverage online (see Anderson et al., 2010). It appears that stories relatively ignored or downplayed by the mainstream media are gaining greater exposure online and that: 'online media are providing different and new portrayals of issues rather than amplifying' traditional news media (Cacciatore et al., 2012: 1051). However, this does not necessarily mean that this is evidence of a more rounded public sphere. While people may have the opportunity of accessing a greater diversity of sources and varied viewpoints, evidence suggests that in practice most individuals rely on cognitive shortcuts and rarely move beyond the first page of the search results (Pan et al., 2007). As discussed in Chapter 5, with reference to the BP oil disaster, search engine results can be manipulated. Moreover, according to the selective exposure theory we tend to actively seek out content that fits in with our viewpoint (Iyengar & Hahn, 2009) and there is a trend for increasingly personalised news feeds that adds a further layer of filtering (Turow, 2012). Twitter reaches a relatively small section of the population who are likely to already have interests on a given topic by virtue of receiving a feed of regular tweets on science and technology issues, and regular Twitter users are likely to be young, urban-dwellers and college-educated. Finally, while blogs provide the opportunity for citizens to air their views few members of the public regularly access blogs about science and technology issues, and they tend to be strongly partisan (see Gavin, 2009; Schäfer, 2012). For example, a representative UK survey by IPSOS-MORI/BIS (2011) found that only 2 per cent of respondents claimed to regularly get their information about scientific debates from science blogs. Nevertheless 19 per cent claimed to gain their information about science from other internet websites so it appears that online sources more generally are becoming a more significant source of news.

As mentioned above, a small but growing number of studies have examined media representations of synthetic biology, but few have examined online media coverage. At this stage in the debate we can identify many similarities with the previous reporting of gene technology that, as mentioned before, was largely positive. A similar type of language seems to be being deployed and there are high expectations. Gschmeidler and Seiringer (2012) analysed German language articles in Austrian, German and Swiss print media (daily press and magazines) from 2004–2009, including online versions, popular science newsletters and journals, and online media (including portals of selected broadcast stations presenting summaries of radio or television broadcasts). Levels of coverage significantly increased since 2004, peaking in 2008. However, a third of articles reporting on the field did not specifically use the term 'synthetic biology'; indeed, much of it was found to overlap with general biotechnology coverage. The majority of articles (83 per cent) emphasised benefits – especially for energy generation and environmental applications – while only 51 per cent of articles emphasised risks, such as biosecurity or biosafety issues. The majority of articles (69 per cent) were published in a specialist science/technology section suggesting that the issues have yet to be perceived as having wider interest.

The final section of the chapter discusses the views of scientists and journalists about the coverage of emerging technologies.

Scientists' and journalists' views on coverage of emerging technologies

As alluded to earlier, the coverage of emerging technologies tends to be strongly event-centred rather than issue-driven. News articles about emerging technologies tend to hinge around an event such as a working party being set up, a press conference being organised or a dramatic new breakthrough occurring. One Science Editor for a national leading newspaper remarked: 'Nearly all the articles I've written have been in response to an event.' Routine stories often need a hook, such as the intervention of a prominent scientist or celebrity figure. Journalists frequently comment on the difficulties of getting nanotechnology related stories published by their newspapers. An environmental correspondent describes a cycle of coverage whereby the intervention of a member of the Royal family broke the story in the national press, but interest quickly subsided:

> It's quite hard to break into those circles...it can be an intervention, like Prince Charles' intervention when he wrote for the *Independent on Sunday*. That broke the cycle and brought it up a bit, but it didn't really take off...something has to happen to break into the public domain in a big way and it has to be kept going by constant stories.
> (Environmental Correspondent, quality UK newspaper)

As we saw in Chapter 3, journalists tend to view their role as being detached and position themselves as outside of the public engagement debate. In the main science correspondents working on national daily newspapers (at least the ones we interviewed) do not see their job as to interpret the science for the public and take a particular position on it. The following comment was typical of our interviewees' responses:

> Well the role of the media is to report what's going on and to try to report it honestly and that applies to nanoscience as much as anything else. You know, you're not supposed to take the role on yourself

to take a particular point of view on the subject. There are lots of people there to do that but it's not our job.

(Environmental Correspondent, quality UK newspaper)

Similarly, when asked about what role the media should play in the public's dialogue with science another journalist responded:

I'm in the business of selling newspapers, that's it, end of story. You know, there's nothing else to ... I write stuff that's put in the paper that is consistent with the identity of the paper to sell to our target readership, you know ... there is no mission to enlighten or reveal the truth.

(Science Correspondent, quality UK newspaper)

For science or environmental correspondents the key consideration is what is likely to appeal to their specific target audience. News values are paramount. As one science correspondent explained:

I mean at the end of the day the media are trying to sell newspapers or win viewers for television programmes or whatever and we're going to do things that we think are new and interesting as a result of that. We're not in the business of sort of public communication of science. There's a big difference between journalism reporting and public information as it were.

(Science Correspondent, quality UK newspaper)

Furthermore, evidence suggests that they rely heavily on scientists as sources. This comprises of a relatively select band of scientists from university institutes and the research councils with whom they have cultivated a relationship of trust. They tend to be wary of pressure group sources as the following comment illustrates:

I'd rather, you know, ah try and translate from what a scientist tells me what's going on than from, you know, a company or a pressure group. I think in general they tend to be a bit more sort of informed about what the reality is. But I mean when you say what role do they play I mean it's not like scientists are beating a path to my door to talk about it, but I do get a fair amount of press releases ... I've got more time for scientists. I mean there are pressure groups and pressure groups. Some of them are very well informed, very diligent.

It's a shame pressure groups don't do real research, or commission real research, because I think that would give them a bit more oomph.
(Science Correspondent, quality UK newspaper)

Another journalist expressed the same sorts of concerns:

I mean I would generally trust say what I get from the Royal Society more than I would trust what I get from Greenpeace...they have a track record and they deal with...scientists from that sort of background...conforming to the scientific method and doing work that is backed up by evidence rather than opinion. And I guess with a lot of NGOs I don't think they've covered themselves in glory with other campaigns that they've run, so one would treat the campaigns they're running on nanotechnology with a degree of scepticism because of that.
(Science Correspondent, quality UK newspaper)

The journalists we interviewed commented that there had been a noticeable change in the more proactive stances of some scientists involved in the nanotechnologies field, following the GM and BSE crises. For example, one journalist observed:

I think they've learned from things like the GM experience on that and have actually realised that they need to sell their work at an early stage because it does have the potential to scare people.
(Science Editor/Correspondent, quality UK newspaper)

Issues were also raised about non-specialist political reporters covering complex scientific issues. One journalist, for example, commented on the sensationalist reporting of the case involving Dr Arpad Pusztai by *The Guardian* (a scientist whose research on GM potatoes led him to raise concerns about their safety) who was blamed by many for seemingly turning the public against GM food:

you sometimes get disasters like, dare I say it, *The Guardian* and the Pusztai affair handled by general correspondents...who sort of took a naïve, scepticism free line on what they were being told and the story was blown up very big...it's a bit scary to think that in that case, you know, someone has swallowed the Prince's anxieties hook, line and sinker without thinking, hang on, what do the real scientists think about this stuff? Of course, the Prince he positioned the story in a clever way so he could claim that it was all those ridiculous journalists that hyped it up at the end. But that's the sophisticated game that he plays with the media.

(Science Correspondent, quality UK newspaper)

For some of our interviewees the risk of nanoparticles was 'normalised'; it was perceived as a 'typical' risk associated with similar sorts of particles derivative of other domains. Defined within these terms, nanoparticles will not be characterised as posing an 'abnormal risk' (Kasperson et al., 2001). For example, one journalist commented:

It's essentially just a new name for Chemistry . . . Safety protocols need to be redrawn to take account of toxic properties of substances, chemicals that can harm the environment. But in general nanotechnology raises no fresh safety or ethical issues that are different from other aspects of science.

(Science Correspondent, quality UK newspaper)

However, for some of the journalists we interviewed there was recognition of significant uncertainty about the risks associated with nanoparticles: 'You know I think there are significant issues about particularly particulates that we haven't really done enough research to know what the hell's going on there, so that's interesting' (Science Editor, quality UK newspaper).

Indeed, the degree of uncertainty surrounding this provides one possible partial explanation for the lack of media attention to this topic. Thus far there have been few newsworthy 'events' that have triggered extensive news media coverage of nanotechnology in the UK press. In 2004 various UK events and source interventions provided politicised, ideological or 'celebrity' news hooks for coverage. However, this often drew attention to areas which journalists swiftly argued lacked credibility, such as concern around the 'grey goo' scenario of runaway self-replicating nanobots. Furthermore such coverage drew attention away from risks, such as those around nanoparticle safety, which journalists perceived as being more feasible. Nanoparticle safety is an issue of long-term risk and current scientific uncertainty and as such has remained relatively under-explored and attracted little in-depth news media attention. Despite scientific and journalistic awareness that it may be an area which generates future news coverage (see Wilkinson et al., 2007), it is not possible to tell from these findings alone if journalists are avoiding producing unnecessary public alarm or criticism from the scientific community or whether they are just biding their time for more extensive research to develop in the future.

The scientists whom we interviewed expressed general dissatisfaction with press coverage of nanotechnologies (see Petersen et al., 2009). Accuracy was a particular concern including the use of particular imagery such as 'nanobots', misleading metaphors and the blurring of fact and fiction. There was some recognition that hype could sometimes be beneficial, at least in some cases as a form of attention grabbing. Views were mixed and reflected the tension scientists experience in needing to get their work publicised but concerns over potential over-selling:

> Most stories result due to a breakthrough in science, which is reported but this is extrapolated to suit the sensationalism and sell papers. This popularizes the subject but in the wrong way.
>
> (Professor, Technology Institute)

> I definitely share the view that nanotechnology has been hyped out of all proportion, which then means that the actual science cannot live up to the expectations of nanobots etc. However, almost any discussion of science in the media is a good thing.
>
> (Professor of Organic Chemistry)

> The newspaper coverage with which I've been involved ... has generally been well-balanced ... However, 'nanobots' always feature highly in press coverage, as do artists' renditions of Fantastic Voyage-like 'nanosubs' hunting down viruses in the bloodstream ... It seems that such 'sensationalised' images are used to 'spice up' otherwise well-balanced articles. Is this detrimental to science? It's a moot point – if the 'nanobot' or 'nanosub' image succeeds in attracting a reader's attention to a well-balanced and scientifically correct article then one might argue that the 'artist's impression' has served an appropriate purpose and this has been beneficial to science. If, however, the nanobot or nanosub image is the only information that remains with the reader, then this is extremely misleading and is rather detrimental to the future of nanoscience.
>
> (Professor, School of Physics and Astronomy)

While for scientists the role of the media is typically seen as one of public education, for journalists it is to provide information and entertainment. When asked what the role of scientists should be in explaining nanotechnologies to the public the following comment was typical: 'To explain the science facts, not the science fiction' (Professor of Organic Chemistry). However, to imply that the 'facts' should be allowed to speak for themselves suggests a limited understanding of science mediation. Pressures to gain future funding may lead them in some instances to spin or hype their work. Scientists need to develop a greater awareness of the underlying factors influencing the framing of news stories on nanotechnologies, and reflect on their own involvement and assumptions in this process.

Summary

Technological innovations, including those arising from the convergence of technologies, are developing far ahead of public debate about their implications. This raises a number of pressing issues. While there has been resistance among some radical eco-anarchists, nanotechnologies and synthetic biology are still for most people an unknown quantity – a non-issue. However, public dialogues on nanotechnologies and, more recently, synthetic biology suggest a number of underlying public concerns when prompted including the lack of knowledge about long-term impacts, the need for effective regulation, the possibility that new forms of pollution will be released and ethical issues over who benefits and who controls the technology.

   The news media play a potentially key role in framing expectations around emerging technologies, especially where people lack familiarity with them, although there is no simple relationship between public attitudes and media content. Past experiences with emerging technologies has shown that news media coverage has tended to be overwhelmingly positive in the early stages of debate. However, in some instances it can become more conflict-driven if interest groups gain more legitimacy through the regulatory system, the topic becomes more politicised and it moves into a more overtly political arena. The evidence reviewed here suggests that more negative framing of emerging technologies may be developing online compared with mainstream media. Yet these channels appear to be primarily utilised by a very small proportion of consumers who are already more informed about such issues than most citizens. Both synthetic biology and nanotechnology are, at present, largely in the realms of specialist science and technology media, and have not entered popular culture.

   There is currently considerable ambivalence about both nanotechnologies and synthetic biology. This appears to be the case both for lay publics and for many NGOs. Also, different responses to the technologies are likely to develop in different cultural contexts. For example, in Europe the legacy of the GM conflict could colour people's views about the use of nanoparticles in food and food contact materials, perhaps leading them to become stigmatised. Whereas in the US food applications tend to be viewed in a more positive light and synthetic biology may be more likely to generate conflict, given that it may have the potential to divide opinion along strongly held religious/ideological lines. At the present time, however, there are no signs that either nanotechnologies or synthetic biology is likely to engender public controversy in the short term. However, as we have seen the internet has provided new ways of visualising emerging technologies and more negative framings are starting to appear. Whether stigmatisation of nanotechnologies or synthetic biology occurs in the future remains to be seen but trust will be a key issue.

Further reading

Anderson, A., Petersen, A., Wilkinson, C. and Allan, S. (2009) *Nanotechnology, Risk and Communication*. Palgrave Macmillan: Houndmills.

Kasperson, R., Jhaveri, N. and Kasperson, J. X. (2001) 'Stigma and the Social Amplification of Risk: Toward a Framework of Analysis', in J. Flynn, P. Slovic, and H. Kunreuther (eds.) *Risk, Media and Stigma: Understanding Public Challenges to Modern Science and Technology*. London: Earthscan, 9–30.

Priest, S. (ed.) (2012) *Nanotechnology and the Public: Risk Perception and Communication*. New York: Taylor and Francis.

7
Future Directions

The previous chapters highlighted the shifting role of the news media in communicating environmental conflict and emerging technologies in contemporary society. In Western societies many inhabit what Nick Couldry (2012: 55) refers to as a 'supersaturated media landscape'. The sheer volume of information that publics can increasingly access means that we have to engage in continuous filtering processes. Online and offline media environments interact in complex and sometimes contradictory ways. The digital revolution has transformed not only how people in the developed world access and interact with information, but who has access and who produces content. New media actors have emerged and journalists routinely use social networking tools to source information. Alongside this, the frames and agendas of 'mainstream' news media still exert considerable influence; indeed the distinction between 'alternative' and 'traditional' media is increasingly blurred.

At the beginning of this book the work of social theorist Ulrich Beck was highlighted as providing one possible lens through which to understand the role of the media in defining a variety of environmental risks and threats that characterise late modernity. His theory was shown to powerfully draw attention to the unprecedented global nature of such risks and the role of the 'relations of definition' – the competing rationality claims and the organisational interests and institutions that assess and manage such risks. However, his theorisation of the media is rather undeveloped, tending to present a static and a-historical view that glosses over shifting power dynamics. Beck tends to treat the media and publics as monolithic and he makes strong claims about media effects without providing empirical evidence. Also, he pays little attention to how different national and political contexts and cultures may influence coverage. Finally, the narrow focus on the media spotlight directs

162

attention away from examining news production processes and the less visible dimensions of power.

By contrast, Castells' illuminating work represents a more fully developed perspective. His theory is less conjectural, more nuanced and gives greater emphasis to the dynamic agency of actors that shifts over time. Castells recognises that the media are diverse and networks are continually reconfiguring around new issues and changing coalitions. However, while he is right to point to a synergy between the fluid, non-hierarchical structure of the internet and that exhibited by a new wave of global activism his account is lacking in a number of respects. He too is relatively silent about the back-stage everyday dynamics of news production. Much more could be said about how routine processes and constraints, cultural assumptions and source relations shape environmental coverage. Ethnographic studies have shown how social dominance alone does not automatically guarantee successful news entry (see Anderson, 1997; Lester, 2007). However, I have argued that challenger groups face myriad hurdles in seeking to frame issues on their terms. For example, while celebrity backing with its associated social capital may elevate the status of a campaign, this is often short-lived and rarely leads to policy change in the short term. Also, mass self-communication is already being co-opted into corporate frames, and information saturation and digital divides limit the extent to which social actors can radically alter the balance of power. While Castells recognises that most people are not part of the core networks that structure contemporary society, he contends that mass self-communication has 'deeply modified the gatekeeping capacity of the programmers of mass communication' (2009: 419). However, in most cases citizens contribute to debates about environmental issues through user-generated comment rather than news reporting (although of course user-generated comment may be indirectly influential if a topic generates sufficient background noise in the system). And while there has undoubtedly been an increase in the number and range of voices competing to make their views heard via digital media some recent studies suggest that online news in general may be as heavily reliant on official sources as offline news (e.g. Curran et al., 2013). There are also a number of potential drawbacks associated with 'clicktivism' or 'slacktivism' discussed in Chapter 2. Social media such as Facebook and Twitter can be powerful tools for environmental campaigners (for example, in online petitions as well as generally increasing the profile of an issue). However, there are important limitations when it comes to mobilising supporters to engage in other forms of protest action, and in developing a strategy and building consensus (see Cox, 2013). Finally there is the question of how far this spills over into influencing wider publics.

The metaphors of 'network society' and 'risk society' are useful up to a point. However, any attempt to comprehensively encapsulate the essence of contemporary society inevitably produces a broad-brush account. We inhabit a complex communications environment with rapid, dynamic, over-lapping flows that reciprocally impact on one another in non-linear ways. It may be tempting to generalise about environmental issues but they may capture or escape (intentionally or unintentionally) the media spotlight at particular points in time for very different reasons. While some environmental issues are more locally based, others have a much more global reach. The scale and complexity of the political, socio-economic and regulatory challenges posed by climate change, for example, is unparalleled (Cox, 2013). Typically environmental news makes up a miniscule proportion of all news content (Project for Improved Environmental Coverage, 2013) and at the same time stories about green issues compete with one another for news media attention. There are many factors that determine the rise and fall of particular environmental issues in the media. Some issues may resonate with the public more than others, some have greater visibility and visual appeal, and some are perceived to have greater immediacy or are more politically opportune. Also, it is easier for issue entrepreneurs to find new angles on certain topics than it is on others. What we tend to find is that as the site of policy debate shifts from administrative to overtly political arenas, and there is a rise in agenda-building activities, media attention spikes (Nisbet et al., 2003). Also, where there is building pressure for regulatory action that is not covered by existing mechanisms, media reporting tends to be especially intense (Torgersen & Hampel, 2012).

It is also relevant to consider how the strategies and goals of environmental NGOs vary and over time targets may change. Many pressure groups focus upon attracting media attention in the early stages and then move on to direct more of their energies to parliamentary activities. They often focus on a range of different campaign areas and pragmatically a number of factors determine which topic becomes the focus at any one point in time. This includes perceived levels of public interest, timeliness, trigger events and media appeal. For example, apart from the Canadian Erosion, Technology and Concentration group (ETC) and FoE few environmental NGOs appear to currently consider synthetic biology to be a major campaign issue. In part this is because regulatory and official policy institutions have taken it upon themselves to consider ethical, legal and social issues (ELSI), previously the territory of campaign organisations, themselves (Torgersen & Hampel, 2012). Similarly, with nanotechnology attempts to 'move public engagement upstream' – the consideration of a technology early on before significant research and development decisions so that decision-makers are informed by public views – may have had the effect of closing down rather than opening up wider discussion. Upstream public engagement activities are top-down participatory projects (citizen juries, focus groups, etc.) that are deliberately organised, often by technology assessment specialists or social scientists, on behalf of government or industry actors. Recognising the diversity of public opinions, science and society programmes has rightly sought to move away from the discredited deficit model. The deficit model rested on the assumption that a disparity in views between scientific experts and some publics simply resulted from a deficit of knowledge on science and technology; all that was needed to bring their views closer into line was greater education. While the impact and influence of the news media on public attitudes to environmental issues (and emerging technologies more generally) is considerably more complex than the deficit model allows, as Nisbet et al. observe:

> Not only do the media influence the attention of competing political actors and the public but the media also powerfully shape how policy issues related to science and technology controversy are defined, symbolized, and ultimately resolved.
>
> (2003: 38)

However, as discussed in Chapter 3, early agenda-setting theories tended to assume a linear causal relationship between the hierarchy of issues on media agendas and that of public agendas. They downplayed the complex web of interacting effects and non-media influences, and glossed over the hidden face of power. I have argued that control over the media is as much to keep certain issues marginalised or hidden as it is to publicise them. In other words, as important as it is to examine which issues receive prominence in the news media, it is crucial to also focus on the ability of news sources to keep issues *off* the news media agenda. This requires a sustained focus on the strategic activity of news sources and clearly poses a methodological challenge. While there are undoubtedly more opportunities for publics to make their voices heard, there is little evidence to suggest that we have entered into an age of minimal news media agenda-setting effects. Rather, we can identify a complex inter-media agenda-setting process whereby social media and traditional media dynamically interact.

The concept of framing sheds more light on the processes of selection and emphasis in news media production, which makes some aspects of a news story more salient than others. I have argued that once an issue is framed in the formative stages of a debate by the news media it can be very hard to shift it to another interpretation. Studies that focus exclusively on examining media representations inevitably produce a partial and narrow picture of what is going on; they shed little light on the wider cultural politics of environmental issues. Ethnographic case studies that trace the evolution of an issue from production, to representation, to consumption produce a much more rounded and useful analysis that is more likely to capture the complexity and contingencies of social processes and the wider play of political power (Anderson, 2009). There is a need for rigorous empirical analysis of source-media relations including more systematic analysis of the strategic and tactical action of news sources (e.g. scientists, NGOs, industry, policymakers) in relation to the media. Gaining media access and achieving coverage is only half the battle. How news sources' claims are framed, and whether they are portrayed as legitimate and credible, is of critical importance.

While the environmental communication field has undoubtedly made great strides over the past two decades there is still a tendency then for studies to be media-centric. This approach is unable to reveal less visible aspects of news production processes and the hidden faces of power. Moreover, it is important to recognise the complexity of source-media relations, given the frequent non-attribution of news sources within news articles. If a news source has little visibility within the news media we cannot simply infer that no framing contest is going on.

Torgersen and Hampel's Gate Resonance Model provides a useful analytical framework that goes beyond a media-centric approach and maps how conflict is played out across a range of different arenas (see Figure 7.1). It builds on Hilgartner and Bosk's earlier Public Arenas Model (see Hilgartner & Bosk, 1988), which represented a move forward from early work tracing the rise and fall of social problems that tended to focus on a limited number of fora, namely the mass media and public agendas (for example, Downs, 1972). Of particular significance here was the observation that different public arenas have different 'carrying capacities' – organisational procedures, mechanisms and channels governing the flow of information – that limit the number of problems that can receive widespread attention. While the role of the media is key the Gate Resonance Model recognises that: 'different arenas vary in importance over time and conflicts transgress from media discourse to regulatory and parliamentary arenas and back again' (Torgersen & Hampel, 2012: 136).

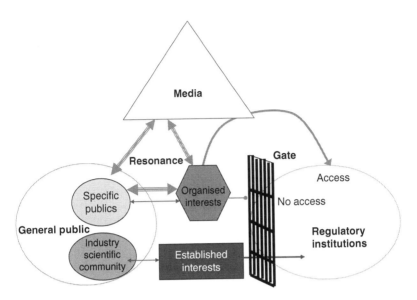

Figure 7.1   The Gate Resonance Model
Source: Reproduced from Torgersen and Hampel 2012: 137.

The Gate Resonance Model identifies a number of different arenas where conflict is played out, including: the media, parliament, regulatory institutions, interest and pressure groups, scientific communities and industry. Instead of being static the model argues that the influence of actors shifts over time. It also recognises the existence of different 'publics' that may overlap with, but are distinct from, those who share similar worldviews. Central to the model is the 'Gate' – the key filter in the regulatory system's processing of various interests (Torgersen & Hampel 2001: 29). Regulatory institutions may be at the national level (for example, the Environment Agency in the UK or the Environmental Protection Agency in the US) or the international level (for example, the United Nations). Such institutions are responsible for processing regulatory needs and selecting issues that fit into formal criteria. Also, they determine which interest groups are included in policy-making debate, and get the opportunity to influence decisions, and those that are excluded. Established interests tend to be involved in defining the Gate so that their arguments are supported. If non-established interests are not represented through the usual policy-making channels then under certain circumstances they may be able to circumvent the gate function and gain access to the regulatory system. However, for this to happen there has to be sufficient issue salience to mobilise publics and attract extensive media attention. This was precisely what happened with GM crops.

Torgersen and Hampel also usefully direct attention to the early warning screening activity undertaken by regulatory institutions. Here 'detectors' actively observe and respond to threats that could challenge the dominant frames. For example, in relation to synthetic biology, ELSI programmes have institutionalised this form of screening activity at an early stage. The role of detectors in regulatory organisations is to scan the media, and observe and assess scientific, societal or political developments that could potentially pose a challenge to dominant frames. This provides a means by which, on occasion, individual civil servants may be able to exercise considerable influence on institutional policy through enabling some arguments to gain back-door access if a substantial weight of evidence is provided (see Torgersen & Hampel, 2001).

The Gate Resonance Model also recognises that the regulatory system is not homogeneous. The regulatory system is made up of Parliament and the government with its body of administration. It comprises different regulatory institutions that work with one another to establish rules, but they have their own differing interests and views, and they compete against each other for power. Regulators have their own agenda that is separate from, but embedded in, the wider political agenda of the government. The model rightly acknowledges that the political arena is far from static and over time actors rise and fall in prominence and influence. Moreover, the media are not perceived as homogeneous. Instead they are seen as constituting multiple arenas addressing different publics and governed by different constraints. They frame environmental issues according to news values (drama, visual appeal, conflict, negativity, human interest) and the climate of public opinion. We might also add that campaigning organisations compete against one another for attention and at different points in time their interests may coalesce or collide.

While this is a very useful analytical framework it has surprisingly little to say about the new communications environment. Past experiences with emerging technologies have shown that mainstream news media coverage has tended to be largely positive in the early stages of debate. However, as we have seen, in some instances it can become more conflict-driven if interest groups gain more legitimacy via the regulatory system. The issue then becomes more politicised and it moves into a more overtly political arena. As discussed in Chapter 6, there is some evidence to suggest that a more negative framing of emerging technologies is developing online compared with mainstream media and this may be increasing over time. Cacciatore et al. (2012), for example, found that there is significantly more coverage of risk and regulation, and more environmentally themed content on nanotechnology online than in the US printed press. The internet thus appears to be providing, to some extent, a means to bypass traditional gatekeepers. Indeed Cacciatore et al. concludes: 'online media are providing different and new portrayals of issues rather than merely amplifying traditional US news media portrayals' (2012: 1051). However, at the present time these channels appear to be mainly utilised by a small minority who are already more informed about such issues than most citizens and it remains to be seen whether they will have wider influence in the future. Just because people have the ability to search for a variety of different perspectives on an issue via the internet does not mean that in practice they do. Furthermore even if publics do gain more exposure to a variety of frames through online exposure this does not necessarily result in a wide variety of perceptions. People tend to graze on media and selectively access material that fits in with their own worldviews. Also, as Cacciatore et al. (2012) point out, online media feeds are becoming increasingly personalised based on consumer's previous search histories and where they are geographically located. Search engines are also open to manipulation as we saw in the case of the BP oil disaster discussed in Chapter 5.

One of the key challenges that we face is getting to grips with the complex new media landscape, and developing conceptual frameworks and methodologies that are sophisticated enough to capture this. Within the increasingly international, multi-digital, interactive and fragmented media environment there is a growing need to examine how competing rationality claims are framed by different media given that they are complexly differentiated and governed by different political, economic and organisational constraints. There is a lack of studies that focus upon the different kinds of effects of online and offline media (see Schäfer, 2012). Social network analysis could shed new light on how flows reciprocally impact on one another in non-linear ways. News rhythms are speeding up and citizen journalism is becoming more important.

Looking ahead to the future, there is also considerable scope to examine in more depth how the various actors involved in environmental communication (e.g. scientists, policymakers, advocacy groups, industry and government) are framing the issues. What role can opinion leaders play in influencing the trajectory of controversies over environmental issues both online and offline? This involves examining strategies based on textual *and* visual material. The study of visual representations of the environment is an important but, until relatively recently, neglected area of study (see Anderson, 2009; Hansen & Machin, 2013). Yet as Lester notes: 'Attempts to harness the power of the image have been a dominant motif of modern environmental politics' (2010: 142). Visualisations of climate change originating from regulatory, political or scientific discourses of environmental NGOs and governments may be appropriated and on occasion inverted by advertisers (Hansen & Machin, 2013; see Linder, 2006). Indeed, studies suggest that the media visualise the environment in increasingly decontextualised ways through generic global, symbolic and iconic images that are often highly romanticised. This tends to lessen their power and disconnect the issues from broader processes such as capitalism and consumerism. Much more remains to be done to develop a political economy of visual representations of the environment that identifies their ideological underpinnings and pays consideration to inequalities of access to resources. As Hansen and Machin argue:

> While studies of visual environmental communication have hinted at the production site and resource access as important factors in analysis of the construction of environmental images, few have provided analysis of how resource access, ownership and organizational affiliation impinge on their production. This is a site that deserves considerably more research attention.
>
> (2013: 163)

Finally if we are to understand how different socio-economic, political and cultural contexts shape the reporting of environmental news, there is a need for more cross-cultural research. Future research could usefully examine the factors accounting for different levels of coverage in different countries, their different emphases and why the views of developing countries tend to be get little attention (Painter, 2007; Shanahan, 2007). There is often a considerable amount of recycling of news stories from European and US sources and more work also remains to be done to further understand the role of global news agencies/wire services. Future research will need to examine in-depth the targeting of media by a variety of actors, as well as unravel complex information flows across countries as media increasingly converge.

Bibliography

Adam, D. (2006) 'Royal Society Tells Exxon: Stop Funding Climate Change Denial', *The Guardian*, 20 September.

Allan, S. (2006) *Online News*. Maidenhead, Berkshire: Open University Press.

Allan, S., Adam, B. and Carter, C. (eds.) (2000) *Environmental Risks and the Media*. London and New York: Routledge.

Allan, S., Anderson, A. and Petersen, A. (2010) 'Framing Risk: Nanotechnologies in the News', *Journal of Risk Research*, 13 (1), 29–44.

Allen, J. (2010) 'How Much is BP Spending on Google Search Ads? Search Engine Watch', 9 June. Available at: http://searchenginewatch.com/article/2050818/How-Much-Is-BP-Spending-On-Google-Search-Ads accessed 12 November 2013.

Åm, T. G. (2011) 'Trust in Nanotechnology? On Trust as Analytical Tool in Social Research on Emerging Technologies', *Nanoethics*, 5 (1), 15–28.

American Academy of Arts and Sciences (2010) *Science and the Media*. Available at: http://amacad.org/pdfs/scienceMedia.pdf accessed 22 September 2013.

Anand, M. and Deepa, N. (2013) 'Understanding Trends and Changes in Media Coverage of Nanotechnology in India', *Journal of Scientometric Research*, 2 (1), 70–73.

Anderson, A. (2014) 'Rethinking Climate Change Communication'. Commentary on Part IV 'Future Directions on Climate Politics Scholarship'. In Boykoff, M. and Crow, D. A. (eds.) *Culture, Politics and Climate Change: How Information Shapes our Common Future*. London: Routledge, 221–225.

Anderson, A. (2013) ' "Together We Can Save the Arctic". Celebrity Advocacy and the Rio Earth Summit 2012', *Celebrity Studies*, 4 (3), 339–352.

Anderson, A. (2011) 'Sources, Media and Modes of Climate Change Communication: The Role of Celebrities', *Wiley Interdisciplinary Reviews: Climate Change*, 2 (4), 535–546.

Anderson, A. (2010) 'Communicating Chemical Risks: Beyond the Risk Society'. In Eriksson, J., Gilek, M. and Ruden, C. (eds.) *Regulating Chemical Risks: Multidisciplinary Perspectives on European and Global Challenges*. London: Springer, 29–44.

Anderson, A. (2009) 'Media, Politics and Climate Change: Towards a New Research Agenda', *Sociology Compass*, 3 (2), 166–182.

Anderson, A. (2007) 'Spinning the Rural Agenda'. In Giarchi, G. (ed.) *Challenging Welfare Issues in the Global Countryside*. Oxford: Blackwell, 722–738.

Anderson, A. (2006) 'The Media and Risk'. In Mythen, G. and Walklate, S. (eds.) *Beyond the Risk Society: Critical Reflections on Risk and Human Security*. Buckingham: Open University Press, 114–131.

Anderson, A. (2002) 'The Media Politics of Oil Spills', *Spill Science and Technology Bulletin*, 7 (5), 7–16.

Anderson, A. (1997) *Media, Culture and the Environment*. London: UCL.

Anderson, A. (1993) 'Source-media Relations: The Production of the Environmental Agenda'. In Hansen, A. (ed.) *The Mass Media and Environmental Issues.* Leicester: Leicester University Press, 51–68.

Anderson, A. (1991) 'Source Strategies and the Communication of Environmental Affairs', *Media, Culture and Society*, 13, 459–476.

Anderson, A. A., Brossard, D. and Scheufele, D. A. (2010) 'The Changing Information Environment for Nanotechnology: Online Audiences and Content', *Journal of Nanoparticle Research*, 12 (4), 1083–1094.

Anderson, A. A., Scheufele, D. A., Brossard, D. and Corley, E. A. (2012) 'The Role of Media and Deference to Scientific Authority in Cultivating Trust in Sources of Information about Nanotechnology', *International Journal of Public Opinion Research*, 24 (2), 225–237.

Anderson, A. and Gaber, I. (1993b) 'The Road from Rio: The Causes of Environmental Antisappointment', *Intermedia*, 21 (6), 27–29.

Anderson, A. and Gaber, I. (1993a) 'The Yellowing of the Greens', *British Journalism Review*, 4, 49–53.

Anderson, A. and Marhadour, A. (2007) 'Slick PR? The Media Politics of the Prestige Oil Spill', *Science Communication*, 29 (1), 96–115.

Anderson, A. and Petersen, A. (2012) 'Nanotechnologies and Trust'. In Candlin, C. N. and Crichton, J. (eds.) *Discourses of Trust*. Basingstoke: Palgrave, 86–104.

Anderson, A. and Petersen, A. (2010) 'Shaping the Ethics of an Emergent Field: Scientists' and Policymakers' Representations of Nanotechnologies,' Special Issue of *International Journal of Technoethics*, 1 (1), 32–44.

Anderson, A., Petersen, A., Wilkinson, C. and Allan, S. (2009) *Nanotechnology, Risk and Communication*. Houndmills: Palgrave Macmillan.

Anderson, A., Petersen, A., Wilkinson, C. and Allan, S. (2005) 'The Framing of Nanotechnologies in the British Newspaper Press', *Science Communication*, 27 (2), 200–220.

Andersson, L. and MacDonald, J. (2010) 'An Overview of BP from a Communications Perspective', 20 September. Available at: http://www.scribd.com/doc/37779786/An-Overview-of-BP-Oil-Crisis-From-a-Comms-Perspective accessed 15 May.

Antilla, L. (2010) 'Self-Censorship and Science: A Geographical Review of Media Coverage of Climate Tipping Points', *Public Understanding of Science*, 19 (2), 240–256.

Antilla, L. (2005) 'Climate of Scepticism: US Newspaper Coverage of the Science of Climate Change', *Global Environmental Change*, 15, 338–352.

Aravosis, J. (2011) 'Flashback: BP Photoshops Fake Photo of Command'. America Blog, 20 April. Available at: http://www.americablog.com/2010/07/bp-photoshops-fake-photo-of-command.html

Arnaldi, S. (2008) 'Converging Technologies in the Italian Daily Press 2002–2006: Preliminary Results of an Ongoing Research Project', *Innovation: The European Journal of Social Science Research*, 21 (1), 87–94.

Arnall, A. H. and Parr, D. (2005) 'Moving the Nanoscience and Technology (NST) Debate Forwards: Short-term Impacts, Long-term Uncertainty and the Social Constitution', *Technology in Society*, 27 (1), 23–28.

Atton, C. (2013) 'Activist Media as Mainstream Model'. In Franklin, B. and Carlson, M. (eds.) *Journalists, Sources and Credibility: New Perspectives*. London: Routledge, 61–72.

Atton, C. (2007) 'A Brief History: The Web and Interactive Media'. In Fountain, A., Dowmunt, T. and Coyer, K. (eds.) *The Alternative Media Handbook*. London: Taylor and Francis, 1–33.

BBSRC (Biotechnology and Biological Sciences Research Council, UK) (2011) *Synthetic Biology Dialogue: Overview*. Available at: http://www.bbsrc.ac.uk/web/FILES/Reviews/synbio_summary-report.pdf

Beck, U. (2009) *World at Risk*. Cambridge: Polity Press.

Beck, U. (2006) *Cosmopolitan Vision*. Cambridge: Polity.

Beck, U. (2002) 'The Silence of Words and Political Dynamics in the World Risk Society', *Logos*, 1.4. Available at: http://logosonline.home.igc.org/beck.htm accessed 19 February 2008.

Beck, U. (1999) *What Is Globalization?* Cambridge: Polity Press.

Beck, U. (1998) *World Risk Society*. Cambridge: Polity Press.

Beck, U. (1996) *The Reinvention of Politics: Rethinking Modernity in the Global Social Order*. Cambridge: Polity Press.

Beck, U. (1995) *Ecological Politics in an Age of Risk*. Cambridge: Polity.

Beck, U. (1994) 'The Reinvention of Politics: Towards a Theory of Reflexive Modernisation'. In Beck, U., Giddens, A. and Lash, S. (eds.) *Reflexive Modernisation: Politics, Tradition and Aesthetics in the Modern Social Order*. Cambridge: Polity, 1–55.

Beck, U. (1992) *Risk Society: Towards a New Modernity*. London: Sage.

Beckhusen, R. (2013) 'In Manifesto, Mexican Eco-Terrorists Declare War on Nanotechnology'. *Wired*, 12 March Available at: http://www.wired.com/dangerroom/2013/03/mexican-ecoterrorism/

Bennett, W. L. and Iyengar, S. (2010) 'The Shifting Foundations of Political Communication: Responding to a Defense of the Media Effects Paradigm', *Journal of Communication*, 60 (1), 35–39.

Bennett, W. L. and Iyengar, S. (2008) 'A New Era of Minimal Effects? The Changing Foundations of Political Communication', *Journal of Communication*, 58 (4), 707–731.

Benoit, W. L. (1995) *Accounts, Excuses, and Apologies: A Theory of Image Restoration Strategies*. Albany: State University of New York.

Berube, D. (2006) *Nano-Hype: The Truth Behind the Nanotechnology Buzz*. Amherst, New York: Prometheus Books.

Bhattachary, D., Pascal Calitz, J. and Hunter, A. (2010) 'Synthetic Biology Dialogue' Report. Available at: http://www.bbsrc.ac.uk/web/FILES/Reviews/1006-synthetic-biology-dialogue.pdf accessed 18 October 2012.

Billett, S. (2010) 'Dividing Climate Change: Global Warming in the Indian Mass Media', *Climatic Change*, 99, 1–2.

Birkland, T. A. and Lawrence, R. G. (2002) 'The Social and Political Meaning of the Exxon Valdez Oil Spill', *Spill Science and Technology Bulletin*, 7 (5), 17–22.

Black, R. (2012) 'BBC Correspondent Speaks on the Difficulty of Reporting on Climate'. Available at: http://oneblueworld.blogspot.co.uk/2012/04/bbc-correspondent-speaks-on-difficulty.html accessed 27 April 2012.

Black, R. (2007) 'BBC Switches off Climate Special', 5 September. Available at: http://news.bbc.co.uk/1/hi/sci/tech/6979596.stm accessed 13 July 2009.

Blizzard, E. (2010) 'BP Attempts to Block Media from Filming Extent of the Oil Spill Disaster', *Huffington Post*, 20 May. Available at: http://www.

huffingtonpost.com/edward-f-blizzard/bp-attempts-to-block-medi_b_583355. html accessed 3 May 2012.

Boholm, M. (2013) 'The Representation of Nano as a Risk in Swedish News Media Coverage', *Journal of Risk Research*, 16 (2), 227–244.

Böl, G. F., Epp, A. and Hertel, R. (eds.) (2010) *Perception of Nanotechnology in Internet-based Discussions: The Risks and Opportunities of Nanotechnology and Nanoproducts: Results of an Online Discourse Analysis*. Berlin: Federal Institute for Risk Assessment.

Booker, C. (2011) *The BBC and Climate Change: A Triple Betrayal*. Global Warming Policy Foundation. Available at: http://www.thegwpf.org/images/stories/gwpf-reports/booker-bbc.pdf accessed 19 November 2013.

Boström, A. and Löfstedt, R. E. (2010) 'Nanotechnology Risk Communication: Past and Prologue', *Risk Analysis*, 30(11), 1645–1662.

Bourne, J. K. (2010) 'Gulf Oil Spill: Is Another Deepwater Disaster Inevitable?', *National Geographic*, October. Available at: http://ngm.nationalgeographic.com/2010/10/gulf-oil-spill/bourne-text/1 accessed 1 December 2013.

Boyce, T. and Lewis, J. (eds.) (2009) *Climate Change and the Media*. Oxford: Peter Lang.

Boykoff, M. T. (2011) *Who Speaks for Climate? Making Sense of Media Reporting on Climate Change*. Cambridge: Cambridge University Press.

Boykoff, M. T. (2010) 'Indian Media Representations of Climate Change in a Threatened Journalistic Ecosystem', *Climatic Change*, 99, 17–25.

Boykoff, M. T., Goodman, M. and Littler, J. (2010) 'Charismatic Megafauna: The Growing Power of Celebrities and Pop Culture in Climate Change Campaigns'. *Environment, Politics and Development*, working paper, Kings College, London.

Boykoff, M. T. (2007b) 'From Convergence to Contention: United States Mass Media Representations of Anthropogenic Climate Change Science', *Transactions of the British Institute of Geographers*, 32 (4), 477–489.

Boykoff, M. T. (2007a) 'Flogging a Dead Norm? Newspaper Coverage of Anthropogenic Climate Change in the USA and UK from 2003–2006,' *Area*, 39 (4), 470–481.

Boykoff, M. T. and Boykoff, J. M. (2004) 'Balance as Bias: Global Warming and the US Prestige Press', *Global Environmental Change*, 14, 125–136.

Boykoff, M. T. and Boykoff, J. M. (2007) 'Climate Change and Journalistic Norms: A Case Study of US Mass-media Coverage', *Geoforum*, 36, 1190–1204.

Boykoff, M. T. and Goodman, M. K. (2009) 'Conspicuous Redemption? Reflections on the Promises and Perils of the 'Celebritization' of Climate Change', *Geoforum*, 40, 395–406.

Boykoff, M. T. and Rajan, R. S. (2007) 'Signals and Noise: Mass Media Coverage of Climate Change in the USA and the UK', *European Molecular Biology Organisation Reports*, 8 (3), 207–211.

Boykoff, M. T. and Roberts, T. (2007) *Climate Change and Human Development – Risk and Vulnerability in a Warming World: Media Coverage of Climate Change – Current Trends, Strengths and Weaknesses*. United Nations Development Program Human Development Reports.

Boykoff, M. T. and Yulsman, T. (2013) 'Political Economy, Media, and Climate Change: Sinews of Modern Life', *Wiley Interdisciplinary Reviews: Climate Change*, 4 (5), 359–371.

Brainard, C. (2006) 'Murdoch Goes Green, and his Empire Follows', *Columbia Journalism Review Daily*, 17 November. Available at: http://www.cjrdaily.org/politics/murdoch_goes_green_and_his_emp.php

Braman, D., Kahan, D. M. and Mandel, G. N. (2009) 'Risk and Culture: Is Synthetic Biology Different?' GW Law Faculty Publications and Other Works. Paper 201. Available at: http://scholarship.law.gwu.edu/faculty_publications/201

Branson, R. (2013) 'The Risk of Doing Nothing', 14 October. Available at: http://www.ecorazzi.com/2013/10/14/richard-branson-on-climate-change/ accessed 7 November 2013.

Broadband Commission (2012) The State of Broadband 2012: Achieving Digital Inclusion for All 'Broadband Commission' September 2012. Available at: http://www.broadbandcommission.org/Documents/bb-annualreport2012.pdf

Brockington, D. (2009) *Celebrity and the Environment: Fame Wealth and Power in Conservation*. London: Zed.

Brossard, D., Shanahan, J. and McComas, K. (2004) 'Are Issue-Cycles Culturally Constructed? A Comparison of French and American Coverage of Global Climate Change', *Mass Communication and Society*, 7 (3), 359–377.

Brulle, R. J., Carmichael, J. and Jenkins, J. C. (2012) 'Shifting Public Opinion on Climate Change: Assessment of Factors Influencing Concern Over Climate Change in the US 2002–2010', *Climatic Change*, 114 (2), 169–188.

Brumfiel, G. (2009) 'Supplanting the Old Media?', *Nature*, 458, 274–277.

Bubela, T. M. and Caulfield, T. A. (2004) 'Do the Print Media "Hype" Genetic Research? A Comparison of Newspaper Stories and Peer-reviewed Research Papers', *Canadian Medical Association Journal*, 170 (9), 1399–1408.

Bubela, T. M., Hagen, G. and Einsiedel, E. (2012) 'Synthetic Biology Confronts Publics and Policy makers: Challenges for Communication, Regulation and Commercialization', *Trends in Biotechnology*, 30 (3), 132–137.

Burke, D. (2003) 'This will be Like no other Debate', *Times Higher Education Supplement*, 21 March.

Burri, V. (2009) 'Coping with Uncertainty: Assessing Nanotechnologies in a Citizen Panel in Switzerland', *Public Understanding of Science*, 18 (5), 498–511.

Cacciatore, M. A., Anderson, A. A., Choi, D-H., Brossard, D., Scheufele, D. A., Liang, X., Ladwig, P. J., Xenos, M. and Dudo, A. (2012) 'Coverage of Emerging Technologies: A Comparison Between Print and Online Media', *New Media and Society*, 14 (6), 1039–1059.

Cacciatore, M. A., Scheufele, D. A. and Corley, E. A. (2011) 'From Enabling Technology to Applications: The Evolution of Risk Perceptions About Nanotechnology', *Public Understanding of Science*, 20 (3), 385–404.

Calvert, J. and Martin, P. (2009) 'The Role of Social Scientists in Synthetic Biology', *EMBO reports*, 10 (3), 201–204.

Capstick, I. (2010) 'Five Digital PR Lessons from BP's Oil Spill Response'. 12 July. Available at: http://www.pbs.org/mediashift/2010/07/5-digital-pr-lessons-from-bps-oil-spill-response193.html accessed 15 May 2012.

Carlson, M. (2009) 'Dueling, Dancing, or Dominating? Journalists and their Sources', *Sociology Compass*, 3(4), 526–542.

Carlson, M. and Franklin, B. (2011) 'Introduction'. In Franklin, B. and Carlson, M. (eds.) *Journalists, Sources and Credibility: New Perspectives*. London: Routledge, 1–15.

Carragee, K. and Roefs, W. (2004) 'The Neglect of Power in Recent Framing Research,' *Journal of Communication*, 54 (2), 214–233.

Carvalho, A. (2007) 'Ideological Cultures and Media Discourses on Scientific Knowledge: Re-reading News on Climate Change,' *Public Understanding of Science* 16, 223–243.

Carvalho, A. and Burgess, J. (2005) 'Cultural Circuits of Climate Change in the UK Broadsheet Newspapers, 1985–2003', *Risk Analysis*, 25 (6), 1457–1470.

Castells, M. (2009) *Communication Power*. Oxford: Oxford University Press.

Castells, M. (2007) 'Communication, Power and Counter-power in the Network Society', *International Journal of Communication*, 1, 238–266.

Castells, M. (ed.) (2005) *The Network Society: A Cross-Cultural Perspective*. Cheltenham: Edward Elgar.

Castells, M. (2004) 'Informationalism, Networks and the Network Society: A Theoretical Blueprint'. In Castells, M. (ed.) *The Network Society: A Cross-Cultural Perspective*. Cheltenham: Edward Elgar, 3–45.

Castells, M. (2000) *The Rise of the Network Society. The Information Age: Economy, Society and Culture Volume 1*. Oxford: Blackwell.

Castells, M. and Cardoso, G. (eds.) (2006) *The Network Society: From Knowledge to Policy*. Washington, D.C: John Hopkins Centre for Transatlantic Relations.

Castells, M., Fernandez-Ardevol, M., Linchuan Qiu, J. and Sey, A. (eds.) (2007) *Mobile Communication and Society: A Global Perspective*. Cambridge, MA: MIT Press.

Caulfield, T. (2005) 'Popular Media, Biotechnology, and the "Cycle of Hype"'. *Houston Journal of Health Law and Policy*, 5 (2), 213–233.

Chand, S. (2009) 'UK Climate Scepticism more Common', *BBC News Online*, 10 September. Accessed 16 February 2011.

Chapman, G., Kumar, K., Fraser, C. and Gaber, I. (1997) *Environmentalism and the Mass Media: The North-South Divide*. London: Routledge.

Choi, J. (2012) 'A Content Analysis of BP's Press Releases Dealing with Crisis', *Public Relations Review*, 38, 422–429.

Ciarallo, J. (2010) 'CBS Denied Access to Shoot Oil Spill: Coast Guard and Contractors Say, "This is BP's Rules, Not Ours"'. *PRNewser*, 20 May. Available at: http://www.mediabistro.com/prnewser/cbs-denied-access-to-shoot-oil-spill-coast-guard-and-contractors-say-this-is-bps-rules-not-ours_b3711 accessed 3 May 2012.

CNN (2010) 'Oil Disaster by the Numbers'. Available at: http://edition.cnn.com/SPECIALS/2010/gulf.coast.oil.spill/interactive/numbers.interactive/index.html accessed 11 May 2012.

Cobb, M. D. and Macoubrie, J. (2004) 'Public Perceptions About Nanotechnology: Risks, Benefits and Trust', *Journal of Nanoparticle Research*, 6 (4), 395–405.

Cobb, R. W. and Elder, C. D. (1972) *Participation in American Politics: The Dynamics of Agenda-building*. Boston: Allyn and Bacon, Inc.

Cohen, B. C. (1963) *The Press and Foreign Policy*. Princeton, NJ: Princeton University Press.

Coleman, R. and McCombs, M. (2007) 'The Young and Agenda-less? Exploring Age-related Differences in Agenda Setting on the Youngest Generation, Baby Boomers, and the Civic Generation', *Journalism and Mass Communication Quarterly*, 84 (3), 495–508.

Colom, A. and Pradhan, S. (2013) 'NEPAL: How the People of Nepal Live with Climate Change and what Communication can do'. *Climate Asia*. Available at: http://downloads.bbc.co.uk/rmhttp/mediaaction/pdf/climateasia/reports/ClimateAsia_NepalReport.pdf accessed 27 October 2013.

Conrad, P. (2001) 'Genetic Optimism: Framing Genes and Mental Illness in the News', *Culture, Medicine and Psychiatry*, 25, 225–247.

Conway, M. and Patterson, J. R. (2008) 'Today's Top Story? An Agenda-Setting and Recall Experiment Involving Internet and Television News', *Southwestern Mass Communication*, 24 (1), 31–48.

Corley, E. A. and Scheufele, D. A. (2010) 'Outreach Gone Wrong? When we Talk Nano to the Public, we are Leaving Behind Key Audiences', *The Scientist*, 24 (1), 22.

Corner, A. (2011) 'Hidden Heat: Communicating Climate Change in Uganda, Challenges and Opportunities'. Panos Eastern Africa. Available at: http://psych.cf.ac.uk/understandingrisk/docs/hidden_heat.pdf accessed 16 November 2013.

Cottle, S. (2013) 'Environmental Conflict in a Global, Media Age.' In Cottle, S. and Lester, L. (eds.) *Environmental Conflict and the Media*. Oxford: Peter Lang, 19–36.

Cottle, S. (2011) 'Afterword: Media and the Arab Uprisings of 2011' In Cottle, S. and Lester, L. (eds.) *Transnational Protests and the Media*. Oxford: Peter Lang, 293–204.

Cottle, S. (2010) 'Rethinking News Access', *Journalism Studies*, 1, 427–448.

Cottle, S. (2009) *Global Crisis Reporting: Journalism in the Global Age*. Milton Keynes: OpenUP.

Cottle, S. (2000) 'TV News, Lay Voices and the Visualisation of Environmental Risks'. In Allan, S., Adam, B. and Carter, C. (eds.) *Environmental Risks and the Media*. London: Routledge, 29–44.

Cottle, S. (1998) 'Ulrich Beck, "Risk Society" and the Media: A Catastrophic View?', *European Journal of Communication*, 13 (1), 5–32.

Cottle, S. (1993) 'Mediating the Environment: Modalities of TV News'. In Hansen, A. (ed.) *The Mass Media and Environmental Issues*. Leicester: Leicester University Press, 107–133.

Couldry, N. (2012) *Media, Society, World: Social Theory and Digital Media Practice*. Cambridge: Polity.

Couldry, N. and Curran, J. (2003) 'The Paradox of Media Power'. In Couldry, N. and Curran, J. (eds.) *Contesting Media Power: Alternative Media in a Networked World*. London: Rowman and Littlefield, 3–16.

Cox, R. (2013) *Environmental Communication and the Public Sphere*. London: Sage.

Cox, R. (2006) (First Edition) *Environmental Communication and the Public Sphere*. London: Sage.

Cserer, A. and Seiringer, A. (2009) 'Pictures of Synthetic Biology: A Reflective Discussion of the Representation of Synthetic Biology (SB) in the German-Language Media and by SB Experts', *Systems and Synthetic Biology*, 3 (1), 27–35.

Curran, J., Coen, S., Aalberg, T., Hayashi, K., Jones, P. K., Splendore, S., Papathanassopoulos, S., Rowe, D. and Tiffen, R. (2013) 'Internet Revolution Revisited: A Comparative Study of Online News', *Media, Culture and Society*, 35 (7), 880–897.

Dabrock, P. (2009) 'Playing God? Synthetic Biology as a Theological and Ethical Challenge', *Systems and Synthetic Biology*, 3, 47–54.

Darley, J. (2000) 'Making the Environment News on the Today Programme'. In Smith, J. (ed.) *The Daily Globe: Environmental Change, the Public and the Media*. London: Earthscan, 151–167.

Davies, N. (2008) *Flat Earth News: An Award-Winning Journalist Exposes Falsehood, Distortion and Propaganda in the Global Media*. London: Chatto and Windus.

DeLuca, K. (1999) *Image Politics*. New York: Guilford.

DeLuca, K., Lawson, S. and Sun, Y. (2012) 'Occupy Wall Street on the Public Screens of Social Media: The Many Framings of the Birth of a Protest Movement', *Communication, Culture & Critique*, 5, 483–509.

DeLuca, K. M. and Peeples, J. (2002) 'From Public Sphere to Public Screen: Democracy, Activism, and the "Violence" of Seattle', *Critical Studies in Media Communication*, 19 (2), 125–151.

Ding D., Maibach, E. W., Zhao, X., Roser-Renouf, C. and Leiserowitz, A. (2011) 'Support for Climate Policy and Societal Action are Linked to Perceptions about Scientific Agreement', *Nature Climate Change*, 1, December, 462–66.

Donald, R. (2013) 'Polling Reveals Public Trusts Scientists most on Climate', *The Carbon Brief*, 2 April. Available at: http://www.carbonbrief.org/blog/2013/04/polling-reveals-public-trusts-scientists-most-on-climate accessed 29 November 2013.

Donk, A., Metag, J., Kohring, M. and Marcinkowski, F. (2012) 'Framing Emerging Technologies: Risk Perceptions of Nanotechnology in the German Press', *Science Communication*, 34 (1), 5–29.

Downing, P. and Ballantyne, J. (2008) *Tipping Point or Turning Point? Social Marketing and Climate Change*. London: Ipsos MORI.

Downs, A. (1972) 'Up and Down with Ecology: The "Issue Attention Cycle"', *Public Interest*, 28, 38–50.

Doyle, J. (2011) *Mediating Climate Change*. Aldershot: Ashgate.

Doyle, J. (2007) 'Picturing the Clima(c)tic: Greenpeace and the Representational Politics of Climate Change Communication', *Science as Culture*, 16(2), 129–150.

Dudo, A., Dunwoody, S. and Scheufele, D. A. (2011) 'The Emergence of Nano News: Tracking Thematic Trends and Changes in US Newspaper Coverage of Nanotechnology', *Journalism and Mass Communication Quarterly*, 88 (1), 55–75.

Dunlap, R. E. and McCright, A. M. (2008) 'A Widening Gap: Republican and Democratic Views on Climate Change.' *Environment*, 50, 26–35.

Dunwoody, S. (1999) 'Scientists, Journalists, and the Meaning of Uncertainty'. In Friedman, S., Dunwoody, S. and Rogers, C. L. (eds.) *Communicating Uncertainty: Media Coverage of New and Controversial Science*. Lawrence Erlbaum: New York, 59–80.

Ebeling, M. F. (2008) 'Mediating Uncertainty: Communicating the Financial Risks of Nanotechnologies', *Science Communication*, 29 (3), 335–361.

Entman, R. M. (2004) *Projections of Power: Framing News, Public Opinion, and US Foreign Policy*. Chicago: University of Chicago Press.

Entman, R. M. (1993) 'Framing: Towards Clarification of a Fractured Paradigm', *Journal of Communication*, 43 (4), 51–58.

Entman, R. M. (1991) 'Symposium Framing U.S. Coverage of International News: Contrasts in Narratives of the KAL and Iran Air Incidents', *Journal of Communication*, 41(4), 6–27.

Ericson, R. V., Baranek, P. M. and Chan, J. B. (1991) *Representing Order: Crime, Law and Justice in the News Media*. Milton Keynes: Open University Press.

Fahn, J. (2012) 'Rio+20 Side Events Become the Main Event – Does the Summit Deserve the Scorn and Indifference that it has Received from the Media?', *Columbia Journalism Review*, 22 June.

Fernandez, J. (2010) 'Greenpeace uses National Ads to Attack BP', *Marketing Week*, 20 May. Available at: http://www.marketingweek.co.uk/greenpeace-uses-national-ads-to-attack-bp/3013708.article   accessed   5   November 2013.

Fiscal Times (2010) 'BP Using Google to Manipulate Public Opinion', *Fiscal Times*, 2 June. http://www.thefiscaltimes.com/Articles/2010/06/02/BP-Admits-to-Buying-Oil-Spill-Search-Terms.aspx#page1 retrieved 3 May 2012.

Fischer, D. (2013) 'Climate Coverage, Dominated by Weird Weather, Falls Further in 2012', *The Daily Climate*, 2 January. Available at: http://www.dailyclimate.org/tdc-newsroom/2013/01/2012-climate-change-reporting accessed 1 December 2013.

Fischer, D. (2011) '2010 in Review: The Year Climate Coverage "Fell off the Map"', *The Daily Climate*, 3 January. Available at: http://wwwp.dailyclimate.org/tdc-newsroom/2011/01/climate-coverage accessed 22 March 2011.

Fitzgerald, S. and Rubin, B. (2010) 'Risk Society, Media and Power: The Case of Nanotechnology', *Sociological Spectrum*, 30 (4), 367–402.

Freeman, E. (no date) 'BP, the Gulf Oil Spill and Business Media'. Available at: http://www.collegebizjournalism.org/node/182 accessed 11 May 2012.

Frewer, L. J., Scholderer, J. and Bredahl, L. (2003) 'Communicating About the Risks and Benefits of Genetically Modified Foods: The Mediating Role of Trust', *Risk Analysis*, 23 (6), 1117–1133.

Friedman, S. M. and Egolf, B. P. (2011) 'A Longitudinal Study of Newspaper and Wire Service Coverage of Nanotechnology Risks', *Risk Analysis*, 31 (11), 1701–1717.

Friedman, S. and Egolf, B. (2005) 'Nanotechnology: Risks and Media', *Technology and Society Magazine*, 24 (4), 5–11.

Friends of the Earth (2012) 'Principles for the Oversight of Synthetic Biology'. Available at: http://libcloud.s3.amazonaws.com/93/ae/9/2287/2/Principles_for_the_oversight_of_synthetic_biology.pdf accessed 3 October 2013.

Frohne, L. and Dearing, J. (2010) 'BP Oil Spill Contract and Letter Cloud Media Access', *Powering a Nation Blog*, 31 May. Available at: http://unc.news21.com/index.php/powering-a-nation-blog/bp-oil-spill-contract-and-letter-cloud-media-access.html accessed 3 May 2012.

Fuchs, C. (2009) 'Some Reflections on Manuel Castells' Book "Communication Power"', *TripleC*, 7 (1), 94–108. Available at: http://www.utwente.nl/gw/mco/bestanden/CastellsCommunicationPowerReview.pdf

Fujita, Y., Yokoyama, H. and Abe, S. (2006) 'Perception of Nanotechnology Among the General Public in Japan – of the NRI Nanotechnology and Society Survey Project', *Asia Pacific Nanotechnology Weekly*, 4, 1–2.

Gallup Poll (2009) 'Increased Number Think Global Warming Is "Exaggerated"', 11 March. Available at: http://www.gallup.com/poll/116590/Increased-Number-Think-Global-Warming-Exaggerated.aspx

Gamson, W. (1985) 'Goffman's Legacy to Political Sociology', *Theory and Society*, 14 (5), 605–22.

Gamson, W. A. and Modigliani, A. (1989) 'Media Discourse and Public Opinion on Nuclear Power: A Constructionist Approach', *American Journal of Sociology*, 95 (1), 1–37.

Gamson, W. A. and Meyer, D. (1996) 'Framing Political Opportunity'. In McAdam, D. McCarthy, J. and Zald, M. (eds.) *Comparative Perspectives on Social Movements*. Cambridge: Cambridge University Press, 275–290.

Gamson, W. A. and Modigliani, A. (1987) 'The Changing Culture of Affirmative Action.' In Braungart, R. D. (Ed.) *Research in Political Sociology* (Vol. 3, pp.137–177). Greenwich, CT: JAI Press.

Gamson, W. A. and Wolfsfeld, G. (1993) 'Media and Movements as Interacting Systems', *Annals of the American Academy of Political and Social Science*, 528: 114–25.

Gandy, O. H. (1982) *Beyond Agenda Setting: Information Subsidies and Public Policy*. Norwood, NJ: Ablex.

Gans, H. (1979) *Deciding What's News*. New York, NY: Pantheon.

Gaskell, G., Stares, S., Allansdottir, A., Allum, N., Castro, P. and Jackson, J. (2010) *Europeans and Biotechnology in 2010. Winds of Change? A Report to the European Commission's Directorate-General for Research*. Publications Office of the European Union, Luxembourg. Available at: http://ec.europa.eu/public_opinion/archives/ebs/ebs_341_winds_en.pdf accessed 24 January 2012.

Gavin, N. T. (2009) 'The Web and Climate Change: Lessons from Britain?' In Boyce, T. and Lewis, J. (eds.) *Climate Change and the Media*. Oxford: Peter Lang, 129–142.

Gavin, N. T. (2007) 'Global Warming and the British Press: The Emergence of an Issue and its Political Implications'. Unpublished paper presented at the Political Studies Association's 'Elections, Public Opinion and Parties' Conference, University of West England, Bristol, September 2007.

Gelbspan, R. (2005) 'Snowed: Why is the US Media Silent on Global Warming?' Mother Jones online publication. http://www.motherjones.com/cgi-bin/print_article.pl?url=http://www.motherjones.com/news/feature/2005/05/snowed.html

Gelbspan, R. (2004) *Boiling Point*. Perseus: New York.

Gelbspan, R. (1995) 'The Heat is On: The Warming of the World's Climate Sparks a Blaze of Denial', *Harpers Magazine*, December.

Gerbaudo, P. (2012) *Tweets From the Streets: Social Media and Contemporary Activism*. London: Pluto Press.

Gitlin, T. (1980) *The Whole World is Watching: Mass Media in the Making and Unmaking of the New Left*. Berkeley: University of California Press.

Gladwell, M. (2010) 'Small Change: Why the Revolution will not be Tweeted'. *The New Yorker*, 4 October. Available at: http://www.newyorker.com/reporting/2010/10/04/101004fa_fact_gladwell

Goffman, E. (1974) *Frame Analysis*. Cambridge: Harvard University Press.

Goidel, R., Kirzinger, A., Dunaway, J. and Madison, P. (2012) 'The 2010 Gulf Oil Spill: A Very Local Story', Paper presented at Annual Meeting of the State Politics and Policy Conference, Houston, Texas, 16–18 February.

Goldenberg, S. (2013) 'Reuters Climate-Change Coverage Fell by nearly 50% with Skeptic as Editor', 26 July. Available at: http://www.theguardian.com/environment/2013/jul/26/reuters-climate-change-scepticism-coverage accessed 19 November 2013.

Goldenberg, S. (2010) 'Fox News Chief Enforced Climate Change Scepticism – Leaked Email', *The Guardian*, 15 December.

Gorss, J. and Lewenstein, B. (2005) 'The Salience of the Small: Nanotechnology Coverage in the American Press, 1986–2004'. Paper presented at the International Communication Association Conference, 27 May, New York.

Green Alliance Policy Insight (2012) 'What People Really Think About the Environment: An Analysis of Public Opinion'. Available at: http://www.green-alliance.org.uk/uploadedFiles/Publications/reports/Green_affordable_Pol_Ins_singles.pdf accessed 21 November 2013.

Greenpeace (2012) 'Arcticready or #Shellfail? Shell's Climate Disaster Gets a Rehearsal'. Media Release Issued 8 June. Available at: http://www.greenpeace.org/usa/en/media-center/news-releases/Arcticready-or-Shellfail-Shells-Climate-Disaster-Gets-a-Rehearsal/ accessed 10 November 2012.

Greenpeace (2011) '1982 – Moratorium Puts an end to Commercial Whaling' Available at: http://www.greenpeace.org/international/en/about/history/Victories-timeline/whaling-moratorium/ accessed 29 November 2013.

Greenpeace (2006) 'Cyberactivism Revolutionises Greenpeace Campaigns' Available at: http://archive.greenpeace.org/cyberstory/cyberactivism.htm accessed 27 January 2009.

Greenpeace (2002) 'Activists Bring Oil Disaster to European Ministers'. Available at: http://greenpeace.org/news/details?item_id=83635

Greenpeace International Archives 'Greenpeace's E-mail, Internet and WWW History'. Available at: http://archive.greenpeace.org/history.html accessed 27 January 2009.

Gregory, R., Flynn, J. and Slovic, P. (2001) 'Technological Stigma'. In Flynn, J., Slovic, P. and Kunreuther, H. (eds.) *Risk, Media and Stigma: Understanding Public Challenges to Modern Science and Technology*. London: Earthscan, 3–8.

Griggs, J. W. (2011) 'BP Gulf of Mexico Oil Spill', *Energy Law Journal*, 32, 57–79.

Groboljsek, B., and Mali, F. (2012) 'Daily Newspapers' Views on Nanotechnology in Slovenia', *Science Communication*, 34 (1), 30–56.

Gschmeidler, B. and Seiringer, A. (2012) ' "Knight in Shining Armour" or "Frankenstein's Creation"? The Coverage of Synthetic Biology in German-Language Media', *Public Understanding of Science*, 21 (2), 163–173.

Guthrie, M. (2010) 'No Capping Spill Story', *Broadcasting and Cable*, 6 July. Available at: http://www.broadcastingcable.com/article/453437-No_Capping_Spill_Story.php accessed 13 November 2013.

Gutteling, J. M., Olofsson, A., Fjaestad, B., Kohring, M., Goerke, A., Bauer, M. W. and Rusanen, T. (2002) 'Media Coverage 1973–1996: Trends and Dynamics'. In Bauer, M. W. and Gaskell, G. (eds.) *Biotechnology: The Making of a Global Controversy*. Cambridge: Cambridge University Press, 95–126.

Hall, S., Critcher, C., Jefferson, T., Clarke, J. and Roberts, B. (2013) *Policing the Crisis: Mugging, the State and Law and Order* (2nd ed.). Basingstoke: Palgrave Macmillan.

Handy, R. D. and Shaw, B. J. (2007) 'Toxic Effects of Nanoparticles and Nanomaterials: Implications for Public Health, Risk Assessment and the Public Perception of Nanotechnology', *Health, Risk and Society*, 9, 125–144.

Hannigan, J. (2006) *Environmental Sociology* (2nd ed.). London: Routledge

Hansen, A. (2010) *Environment, Media and Communication*. London: Routledge.

Hansen, A. (2000) 'Claimsmaking and Framing in British Newspaper Coverage of the "Brent Spar" Controversy.' In Allan, S., Adam, B. and Carter, C. (eds.) *Environmental Risks and the Media.* London: Routledge, 55–72.

Hansen, A. (1991) 'The Media and the Social Construction of the Environment', *Media, Culture and Society,* 13, 443–458.

Hansen, A. (1990) *The News Construction of the Environment: A Comparison of British and Danish Television News.* Leicester: Centre for Mass Communications Research, University of Leicester.

Hansen, A. and Machin, D. (2013) 'Researching Visual Environmental Communication', *Environmental Communication,* 7 (2), 151–168.

Harlow, W. F., Brantley, B. C. and Harlow, R. M. (2011) 'BP Initial Image Repair Strategies after the Deepwater Horizon Spill', *Public Relations Review,* 37 (1), 80–83.

Harris Poll (2012) 'Nanotechnology Awareness May be Low, But Opinions are Strong'. Available at: http://www.harrisinteractive.com/NewsRoom/HarrisPolls/tabid/447/ctl/ReadCustom%20Default/mid/1508/ArticleId/1073/Default.aspx accessed 10 September 2013.

Hart Research Associates (2013) *Awareness and Impressions of Synthetic Biology: A Report of Findings.* Washington: Woodrow Wilson International Center for Scholars. Available at: http://www.wilsoncenter.org/sites/default/files/synbiosurvey2013_0.pdf accessed 8 September 2013.

Hart Research Associates, Inc (2008) *Nanotechnology, Synthetic Biology and Public Opinion. A Report of Findings Conducted on Behalf of Project on Emerging Nanotechnologies,* The Woodrow Wilson International Center for Scholars. Available at Available at: http://www.nanotechproject.org/publications/archive/8286/ accessed 30 January 2012.

Harvey, F. (2013) 'BBC Coverage Criticised for Favouring Climate Change Sceptics', *The Guardian,* 28 October.

Herman, E. and Chomsky, N. (2002) *Manufacturing Consent: The Political Economy of the Mass Media.* New York: Pantheon Books.

Hilgartner, S. and Bosk, C. L. (1988) 'The Rise and Fall of Social Problems: A Public Arenas Model', *American Journal of Sociology,* 94, 53–78.

Hoel, A. H. (1986) *The International Whaling Commission 1972–1984: New Members, New Concerns.* Lysaker, Norway: The Fridtjof Nansen Institute.

Hoffbauer, A. (2011) *Beyond the Deepwater Horizon Explosion: What Shaped the Social And Political Engagement of the BP Oil Spill?* Unpublished MA Thesis, Dalhousie University Halifax, Nova Scotia.

Holbert, R. L., Garrett, R. K. and Gleason, L. S. (2010) 'New Era of Minimal Effects? A Response to Bennett and Iyengar', *Journal of Communication,* 60 (1), 15–34.

Hornig, S. (1993) 'Reading Risk: Public Response to Print Media Accounts of Technological Risk', *Public Understanding of Science,* 2, 95–109.

Hornig Priest, S. (2008) 'North American Audiences for News of Emerging Technologies: Canadian and US Responses to Bio- and Nanotechnologies', *Journal of Risk Research,* 11 (7), 877–889.

Horrigan, J. (2006) 'The Internet as a Resource for News and Information About Science'. Pew Internet and American Life Project. Pew Research Center, Washington D.C.

Hough, A. (2010) 'BP Admits it "Photo-Shopped" Official Images as Oil Spill "Cut and Paste" Row Escalates', *The Daily Telegraph,* 22 June.

Available at: http://www.telegraph.co.uk/earth/energy/oil/7904221/BP-admits-it-Photoshopped-official-images-as-oil-spill-cut-and-paste-row-escalates.html

Huffington Post (2010) 'BP, Coast Guard Officers Block Journalists From Filming Oil-Covered Beach', 19 May. Available at: http://www.huffingtonpost.com/2010/05/19/bp-coast-guard-officers-b_n_581779.html

Hutchins, B. and Lester, L. (2006) 'Environmental Protest and Tap-Dancing with the Media in the Information Age', Media, Culture and Society, 28 (3), 433–451.

Ingram, H., Brinton Milward, H. and Laird, W. (1992) 'Scientists and Agenda Setting: Advocacy and Global Warming'. In Waterstone, M. (ed.) Risk and Society: The Interaction of Science, Technology and Public Policy. Dordrecht: Kluwer, 33–53.

International Tank Owners Pollution Federation (2009) 'Oil Tanker Spill Statistics 2008'. Available at: http://www.itopf.com/information-services/data-and-statistics/statistics/documents/Statpack2008_001.pdf accessed 16 October 2009.

IPCC (2013) Working Group 1 Contribution to the IPCC Fifth Assessment Report Climate Change 2013: The Physical Science Basis. Available at: http://www.ipcc.ch/report/ar5/wg1/ accessed 23 November 2013.

IPSOS-MORI (2012) 'Climate Week Poll on Public Attitudes Regarding Climate Change'. Available at: http://www.ipsos-mori.com/Assets/Docs/Polls/ipsos-mori-climate-week-topline-2012.pdf accessed 7 November 2013.

IPSOS-MORI/BIS (2011) Public Attitudes to Science 2011. Available at: http://www.ipsos-mori.com/Assets/Docs/Polls/sri-pas-2011-main-report.pdf accessed 3 October 2013.

IRGC (2010) 'Guidelines for the Appropriate Risk Governance of Synthetic Biology'. Available at: http://www.irgc.org/IMG/pdf/irgc_SB_final_07jan_web.pdf accessed 18 October 2012.

Iyengar, S. (1991) Is Anyone Responsible? How Television Frames Political Issues. Chicago: University of Chicago Press.

Iyengar, S. and Hahn, J. (2009) 'Red Media, Blue Media: Evidence of Ideological Selectivity in Media Use', Journal of Communication, 59 (1), 19–39.

Jacques, P. J., Dunlap, R. E. and Freeman, M. (2008) 'The Organisation of Denial: Conservative Think Tanks and Environmental Scepticism', Environmental Politics, 17 (3), 349–385.

Jiménez-Aleixandre, M. P., Federico-Agraso, M. and Eirexas-Santamaria, F. (2004) 'The Scientific Community as a Source of Information about the Prestige', PCST International Conference. Available at: http://www.consellodacultura.org/mediateca/pubs.pdf/comunicacion_ambiente.pdf

Johnson, B. (2009) 'Astroturfing: A Question of Trust', The Guardian, 7 September.

Jones, J. M. (2011) 'In U.S., Concerns About Global Warming Stable at Lower Levels'. Gallup Politics. Available at: http://www.gallup.com/poll/146606/concerns-global-warming-stable-lower-levels.aspx

Jones, R. (2008) 'The Economy of Promises', Nature Nanotechnology, 3, 65–66.

Jones, S. (2011) 'BBC Trust Review of Impartiality and Accuracy of the BBC's Coverage of Science'. BBC Trust. Available at: http://www.bbc.co.uk/bbctrust/assets/files/pdf/our_work/science_impartiality/science_impartiality.pdf accessed 19 November 2013.

Juhasz, A. (2011) Black Tide: The Devastating Impact of the Gulf Oil Spill. Hoboken, NJ: John Wiley.

Jurgenson, N. (2012) 'When Atoms Meet Bits: Social Media, the Mobile Web and Augmented Revolution', *Future Internet*, 4 (1), 83–91.

Kahan, D. M., Braman, D., Slovic, P., Gastil, J. and Cohen, G. (2009) 'Cultural Cognition of the Risks and Benefits of Nanotechnology'. *Nature Nanotechnology*, 4 (2), 87–90.

Kahan, D. M., Slovic, P., Braman, D., Gastil, J., Cohen, G. and Kysar, D. (2008) 'Biased Assimilation, Polarization, and Cultural Credibility: An Experimental Study of Nanotechnology Risk Perceptions'. Project on Emerging Nanotechnologies, Research Brief No. 3.

Kanerva, M. (2009) 'Assessing Risk Discourses: Nano S&T in the Global South', (UNUMERIT) Working Paper Series No. 063. Available at: http://arno.unimaas. nl/show.cgi?fid=17542.

Kasperson, R., Jhaveri, N. and Kasperson, J. X. (2001) 'Stigma and the Social Amplification of Risk: Toward a Framework of Analysis'. In Flynn, J., Slovic, P. and Kunreuther, H. (eds.) *Risk, Media and Stigma: Understanding Public Challenges to Modern Science and Technology*. London: Earthscan, 9–30.

Katz, I. (2009) 'How the Climate Change Global Editorial Project Came About', *The Guardian*, 9 December. Available at: http://www.guardian.co.uk/ environment/2009/dec/06/climate-change-leader-editorial

Kearnes, M., Macnaghten, P. and Wilsdon, J. (2006) *Governing at the Nanoscale: People, Policies and Emerging Technologies*. London: Demos.

Kellner, D. (2003) *Media Spectacle*. London: Routledge.

Kishimoto, A., Takai, T. and Wakamatsu, H. (2010) 'Public Perceptions of Nanotechnologies in Japan from 2005 to 2009'. *National Institute of Advanced Industrial Science and Technology* (AIST) Available at: http://staff. aist.go.jp/kishimoto-atsuo/nano/nano_20052009_e.pdf accessed 1 October 2013.

Kitzinger, J. (1999) 'Researching Risk and the Media', *Health, Risk and Society*, 1(1), 55–69.

Kitzinger, J., Henderson, L., Smart, A. and Eldridge, J. (2003) 'Media Coverage of the Ethical and Social Implications of Human Genetic Research'. Final Report for the Wellcome Trust (February 2003) Award no: GR058105MA.

Kjærgaard, R. S. (2010) 'Making a Small Country Count: Nanotechnology in Danish Newspapers from 1996 to 2006', *Public Understanding of Science*, 19, 80–97.

Kjølberg, K. L. (2009) 'Representations of Nanotechnology in Norwegian Newspapers – Implications for Public Participation', *NanoEthics*, 3 (1), 61–72.

Kohut, A. (2013) 'Pew Surveys of Audience Habits suggest Perilous Future for News'. Available at: http://www.poynter.org/latest-news/top-stories/225139/ pew-surveys-of-audience-habits-suggest-perilous-future-for-news/   accessed 17 September 2013.

Korosec, K. (2010) 'BP and the Gulf Oil Spill: Misadventures in Photoshop' [Blog commentary], 23 July. Available at: http://www.bnet.com/blog/clean-energy/bp-and-the-gulf-oil-spill-misadventures-in-photoshop/2091   accessed 11 May 2012.

Krauss, C. and Schwartz, J. (2012) 'BP will Plead Guilty and Pay Over $4 Billion', *New York Times*, 15 November.

Kronberger, N. (2012) 'Synthetic Biology: Taking a Look at a Field in the Making', *Public Understanding of Science*, 21 (2), 130–133.

Kronberger, N., Holtz, P., Kerbe, W., Strasser, E. and Wagner, W. (2009) 'Communicating Synthetic Biology: From the Lab via the Media to the Broader Public', *Systems and Synthetic Biology*, 3 (1), 19–26.

Kronberger, N., Holtz, P. and Wagner, W. (2012) 'Consequences of Media Information Uptake and Deliberation: Focus Groups' Symbolic Coping with Synthetic Biology', *Public Understanding of Science*, 21 (2), 174–187.

Krosnick, J. (2010) 'Majority of Americans Still Believe that Global Warming Is Real'. Woods Institute for the Environment, Stanford University, March 9 (unpaginated). Available at: http://woods.stanford.edu/research/majority-believe-global-warming.html

Lakoff, G. (2010) 'Why it Matters How we Frame the Environment', *Environmental Communication*, 4 (1), 70–81.

Lash, S. and Urry, J. (1994) *Economies of Signs and Space*. London: Sage.

Lean, G. (2009) 'Will Rupert Murdoch's Fox News Go for the Kill on Climate Change?', *The Daily Telegraph*, 11 September.

Lean, G. (1995) 'The Role and Responsibility of the Media, Religion, Science and the Environment Conference', *Aegean Sea*, 20–27 September. Available at: http://www.rsesymposia.org/themedia/File/1151633311-Role_Resp_of_Media.pdf accessed 18 November 2013.

Learmonth, M. (2010) 'What Big Brands Are Spending on Google', *Advertising Age*, 6 September. Available at: http://adage.com/article/digital/big-brands-spending-google/145720/ accessed 12 November 2013.

Lee, J. (2010) 'BP, Crisis Communications and Social Media', 1 July. Available at: http://www.bruceclay.com/blog/2010/07/bp-crisis-communications-and-social-media/ accessed 12 November 2013.

Leggett, J. (2001) *The Carbon War: Global Warming and the End of the Oil Era*. London: Routledge.

Leiserowitz, A. (2010) 'Surveying the Impact of Live Earth on American Public Opinion: A Yale University, Gallup and Clearvision Institute Study'. Available at: http://environment.yale.edu/climate-communication/files/LiveEarthReport.pdf accessed 16 July 2010.

Leiserowitz, A., Feinberg, G., Rosenthal, S., Smith, N., Anderson A., Roser-Renouf, C. and Maibach, E. (2014) *What's In A Name? Global Warming vs. Climate Change*. Yale University and George Mason University. New Haven, CT: Yale Project on Climate Change Communication.

Leiserowitz, A., Maibach, E. and Roser-Renouf, C. (2010a) *Climate Change in the Mind: Americans' Global Warming Beliefs and Attitudes in January 2010*. Yale and George Mason University, New Haven, CT: Yale Project on Climate Change. Available at: http://environment.yale.edu/uploads/AmericansGlobalWarmingBeliefs2010.pdf accessed 16 February 2011.

Leiserowitz, A., Maibach, E., Roser-Renouf, C. and Feinberg, G. (2013) *How Americans Communicate About Global Warming*. Yale University and George Mason University. New Haven, CT: Yale Project on Climate Change Communication. Available at: http://environment.yale.edu/climate-communication/files/Communication-April-2013.pdf accessed 29 November 2013.

Leiserowitz, A., Maibach, E., Roser-Renouf, C., Smith, N. and Dawson, E. (2010b) 'Climategate, Public Opinion and Loss of Trust', Working Paper. New Haven, CT: Yale Project on Climate Change. Available at: http://www.climatechangecommunication.org/ . . . /Climategate_Public%20Opinion_and%20Loss%20of%20Trust(1).pdf accessed 16th February 2011.

Lemańczyk, S. (2013) 'Debate on Nanotechnology in the Swedish Daily Press, 2004–2009', *Innovation*, 26 (4), 344–353.

Lemańczyk, S. (2012) 'Between National Pride and the Scientific Success of "Others": The Case of Polish Press Coverage of Nanotechnology, 2004–2009', *NanoEthics*, 6 (2), 101–115.

Lester, L. (2010) *Media and Environment*. Cambridge: Polity.

Lester, L. (2007) *Giving Ground: Media and Environmental Conflict in Tasmania*. University of Tasmania: Quintus.

Lewenstein, B., Radin, J. and Diels, J. (2007) 'Nanotechnology in the Media: A Preliminary Analysis'. In Roco, M. C. and W. S. Bainbridge (eds.) *Nanotechnology: Societal Implications II: Individual Perspectives*, Dordrecht: Springer, 258–265.

Lewis, J., Williams, A. and Franklin, B. (2008) 'A Compromised Fourth Estate? UK News Journalism, Public Relations and News Sources', *Journalism Studies*, 9 (1), 1–20.

Linder, S. H. (2006) 'Cashing-in on Risk Claims: On the For-Profit Inversion of Signifiers for "Global Warming"', *Social Semiotics*, 16 (1), 103–132.

Littler, J. (2008) '"I Feel your Pain": Cosmopolitan Charity and the Public Fashioning of the Celebrity Soul', *Social Semiotics*, 18 (2), 237–251.

Liu, S. B. (2010) 'Trends in Distributed Curatorial Technology to Manage Data Deluge in a Networked World', *Upgrade: The European Journal for the Informatics Professional*, 11 (4), 18–24.

Lively, E., Conroy, M., Weaver, D. A. and Bimber, B. (2012) 'News Media Frame Novel Technologies in a Familiar Way: Nanotechnology, Applications, and Progress'. In Herr Harthorn, B. and Mohr, J. (eds.) *The Social Life of Nanotechnology*. London: Routledge, 223–240.

Lloyd, G. (2010) 'Climate Debate no Place for Hotheads', *The Australian*, 4 December.

Lyndhurst, B. (2009) *An Evidence Review of Public Attitudes to Emerging Food Technologies*. London: Food Standards Agency. Available at: http://www.food.gov.uk/multimedia/pdfs/emergingfoodtech.pdf accessed 1 October 2013.

Lyndon, M. L. (2012) 'The Environment on the Internet: The Case of the BP Oil Spill', *Elon Law Review*. Available at: http://www.elon.edu/docs/e-web/law/law_review/Issues/Elon_Law_Review_V3_No2_Lyndon.pdf accessed 11 November 2013.

MacNaghten, P. (2006) 'Environment and Risk'. In Walklate, S. and Mythen, G. (eds.) *Beyond the Risk Society: Critical Reflections on Risk and Human Security*. Maidenhead: Open University/McGraw Hill, 132–146.

Manning, P. (2001) *News and News Sources*. London: Sage.

McComas, K. and Shanahan, J. (1999) 'Telling Stories About Global Climate Change', *Communication Research*, 26 (1), 30–57.

McCombs, M. (2005) 'A Look at Agenda-Setting: Past, Present and Future', *Journalism Studies*, 6 (4), 543–557.

McCombs, M. E., Shaw, D. L. and Weaver, D. H. (1997) *Communication and Democracy: Exploring the Intellectual Frontiers in Agenda-Setting Theory*. Mahwah, NJ: Erlbaum.

McCormick, S. (2012) 'Transforming Oil Activism: From Legal Constraints to Evidenciary Opportunity', *Sociology of Crime Law and Deviance*, 17, 113–131.

McCright, A. M. and Dunlap, R. E. (2012) 'The Politicization of Climate Change and Polarization in the American Public's Views of Global Warming, 2001–2010', *The Sociological Quarterly*, 52, 155–194.

McCright, A. M. and Dunlap, R. E. (2003) 'Defeating Kyoto: The Conservative Movement's Impact on U.S. Climate Change Policy', *Social Problems*, 50 (3), 348–373.

McCurdy, P. (2012) 'Social Movements and Mainstream Media', *Sociology Compass*, 6 (3), 244–245.

McKnight, D. (2010) 'A Change in the Climate? The Journalism of Opinion at News Corporation', *Journalism*, 11 (6), 693–706.

McManus, P. (2000) 'Beyond Kyoto? Media Representation of an Environmental Issue', *Australian Geographical Studies*, 38, 306–319.

McQuail D. and Windahl, S. (1993) *Communication Models*. Essex: Longman.

Marea (2005) Available at: http://home.wanadoo.nl/klaasvangorkum/prestige/index.htm accessed 4 May 2005.

Melucci, A. (1981) 'Ten Hypotheses for the Analysis of New Social Movements'. In Pinto, D. (ed.) *Contemporary Italian Sociology*. Cambridge: Cambridge University Press, 173–194.

Merry, M. K. (2014) *Framing Environmental Disaster: Environmental Advocacy and the Deep Horizon Oil Spill*. London: Routledge Chapman and Hall.

Messner, M. and Watson Distaso, M. (2008) 'The Source Cycle: How Traditional Media and Weblogs Use Each Other as Sources', *Journalism Studies*, 9 (3), 447–463.

Metag, J. and Marcinkowski, F. (2014) 'Technophobia Towards Emerging Technologies? A Comparative Analysis of the Media Coverage of Nanotechnology in Austria, Switzerland and Germany', *Journalism*, 15 (4), 463–481.

Meyer, D. S. and Gamson, J. (1995) 'The Challenge of Cultural Elites: Celebrities and Social Movements', *Sociological Inquiry*, 65 (2), 181–206.

Miller, M. and Riechert, B. P. (2000) 'Interest Group Strategies and Journalistic Norms: News Media Framing of Environmental Issues'. In Allan, S., Adam, B. and Carter, C. (eds.) *Environmental Risks and the Media*. London: Routledge, 45–54.

Mitchelstein, E. and Boczkowski, P. J. (2009) 'Between Tradition and Change: A Review of Recent Research on Online News Production', *Journalism*, 10 (5), 562–586.

Molotch, H. and Lester, M. (1975) 'Accidental News: The Great Oil Spill as Local Occurrence and National Event', *American Journal of Sociology*, 81, 235–260.

Monbiot, G. (2007) 'The Editorials Urge Us to Cut Emissions, But the Ads Tell a Very Different Story', *The Guardian*, 14 August.

Monbiot, G. (2006) *Heat: How to Stop the Planet from Burning*. London: Random House.

Mooney, C. (2008) 'The Science Writer's Lament'. Available at: http://scienceprogress.org/2008/10/the-science-writers-lament accessed 22 September 2013.

Morgan, G. (2010) 'BP Buys "Oil Spill" Sponsored Links for Search Engines', *New Scientist*, 7 June. Available at: http://www.newscientist.com/blogs/shortsharpscience/2010/06/bp-turns-to-google-ads-to-save.html accessed 3 May 2012.

Morozov, E. (2012) *The Net Delusion: How Not to Liberate the World*. London: Penguin.

Moser, S. C. and Dilling, L. (2011) 'Communicating Climate Change: Opportunities and Challenges for Closing the Science-action Gap'. In Dryzek, J. S., Norgaard, R. B. and Schlosberg, D. (eds.) *The Oxford Handbook of Climate Change and Society*, Oxford: Oxford University Press, 161–74.

Muralidharan, S., Dillistone, K. and Shin, J. H. (2011) 'The Gulf Coast Oil Spill: Extending the Theory of Image Restoration Discourse to the Realm of Social Media and Beyond Petroleum', *Public Relations Review*, 37, 226–232.

Murdoch, R. (2007) 'Speech on Climate Change, Energy Initiative', New York City, 9 May. Available at: http://www.newscorp.com/energy/full_speech.html accessed 5 November 2012.

Murdock, G., Petts, J. and Horlick-Jones, T. (2003) 'After Amplification: Rethinking the Role of the Media in Risk Communication'. In Pidgeon, N., Kasperson, R. E. and Slovic, P. (eds.) *The Social Amplification of Risk*. Cambridge: Cambridge University Press, 156–178.

Mythen, G. (2007) 'Reappraising the Risk Society Thesis: Telescopic Sight or Myopic Vision?', *Current Sociology*, 55 (6), 793–813.

Mythen, G. (2004) *Ulrich Beck: A Critical Introduction to the Risk Society*. London: Pluto Press.

Nature (2003) 'Don't Believe the Hype', *Editorial*, 424, 17 July, p. 237.

Nelkin, D. (1994) 'Promotional Metaphors and their Popular Appeal', *Public Understanding of Science*, 3 (1), 25–31.

Nelkin, D. (1985) 'Managing Biomedical News', *Social Research*, 52 (3), 625–646.

Newell, P. (2000) *Climate for Change: Non-State Actors and the Global Politics of the Greenhouse*. Cambridge: Cambridge University Press.

Newport, F. (2011) 'Americans' Global Warming Concerns Continue to Drop: Multiple Indicators Show Less Concern, More Feelings that Global Warming is Exaggerated'. *Gallup.com*, 11 March. Available at: http://www.gallup.com/poll/126560/Americans-Global-Warming-Concerns-Continue-Drop.aspx accessed 23 March 2011.

Nielsen (2007) *Global Omnibus Survey*. Oxford: UK.

Nisbet, M. (2011) 'As Unemployment Drops Public Belief in Climate Change Shows Signs of Recovery'. *Climate Shift Report*, 2 December. Available at: http://climateshiftproject.org/2011/12/02/as-unemployment-drops-public-belief-in-climate-change-shows-signs-of-recovery/

Nisbet, M. (2008) *Moving Beyond Gore's Message: A Look Back (and Ahead) at Climate Change Communications*. Committee for Sceptical Inquiry online publication.

Nisbet, M., Brossard, D. and Kroepsch, A. (2003) 'Framing Science: The Stem Cell Controversy in an Age of Press/Politics', *Harvard International Journal of Press/Politics*, 8 (2), 36–70.

Nisbet, M. C. and Huge, M. (2006) 'Attention Cycles and Frames in the Plant Biotechnology Debate: Managing Power and Participation', *The Harvard International Journal of Press/Politics*, 11 (2), 3–40.

Nisbet, M. C. and Lewenstein, B. V. (2002) 'Biotechnology and the American Media: The Policy Process and the Elite Press, 1970 to 1999', *Science Communication*, 23 (4), 359–391.

Nisbet, M. and Myers, T. (2007) 'Twenty Years of Public Opinion About Global Warming', *Public Opinion Quarterly*, 7 (3), 1–27.

Nohrstedt, S. A. (1991) 'The Information Crisis in Sweden after Chernobyl', *Media, Culture & Society*, 13 (4), 477–497.

Nolan, J. M. (2010) 'Consensus and Controversy: Climate Change Frames in Two Australian Newspapers'. *Open Access Theses*. Paper 30. Available at: http://scholarlyrepository.miami.edu/oa_theses/30

Norton, D. W. (2010) 'Constructing "Climategate" and Tracking Chatter in the Age of Web 2.0', *Center for Social Media, School of Communication*, American University, Washington D.C.

Olausson, U. (2009) 'Global Warming – Global Responsibility? Collective Action Frames and the Discourse of Certainty', *Public Understanding of Science*, 18 (4), 421–436.

O'Neill, M. (2010) *A Stormy Forecast: Identifying Trends in Climate Change Reporting*. Reuters Institute. Available at: https://reutersinstitute.politics.ox.ac.uk/fileadmin/documents/Publications/fellows__papers/2009-2010/A_STORMY_FORECAST.pdf accessed 5 November 2012.

O'Neill, S. J., Boykoff, M., Niemeyer, S. and Day, S. A. (2013) 'On the Use of Imagery for Climate Change Engagement', *Global Environmental Change*, 23 (2), 395–572.

O'Reilly, G. (2010) BP Buys Oil-related Search Terms as Attempts Continue to Stem Oil Spill, *PR Week*, 10 June. Available at: http://www.prweek.com/article/1009057/bp-buys-oil-related-search-terms-attempts-continue-stem-oil-spill accessed 6 November 2013.

Painter, J. (2010) *Summoned by Science: Reporting Climate Change at Copenhagen and Beyond*. Reuters Institute, University of Oxford.

Painter, J. (2007) 'All Doom and Gloom? International TV Coverage of the April and May 2007 IPCC Reports', Unpublished manuscript, Environmental Change Institute, University of Oxford, UK.

Pan, B., Hembrooke, H., Joachims, T., Lorigo, L., Gay, G. and Granka, L. (2007) 'In Google we Trust: Users' Decisions on Rank, Position, and Relevance', *Journal of Computer Mediated Communication*, 12 (3), 801–823.

Pan, Z. and Kosicki, G. M. (1993) 'Framing Analysis: An Approach to News Discourse', *Political Communication*, 10, 55–75.

Pauwels, E. (2013) 'Public Understanding of Synthetic Biology', *BioScience*, 63 (2), 79–89.

Pauwels, E. (2009) 'Review of Quantitative and Qualitative Studies on U.S. Public Perceptions of Synthetic Biology', *Systems and Synthetic Biology*, 3 (1), 37–46.

Pauwels, E. and Ifrim, I. (2008) *Trends in American and European Press Coverage of Synthetic Biology* (SYNBIO 1, November 2008) Available at: http://www.synbioproject.org/process/assets/files/5999/synbio1final.pdf

PBS (2010) *Gulf Coast Oil Leak: Oil Leak Widget*. Available at: http://www.pbs.org/newshour/rundown/horizon-oil-spill.html accessed 11 May 2012.

Pearce, F. (2010) *The Climate Files: The Battle for the Truth About Global Warming*. London: Guardian Books.

Petersen, A. (2001) 'Biofantasies: Genetics and Medicine in the Print News Media', *Social Science and Medicine*, 52, 1255–1268.

Petersen, A., Anderson, A., Allan, S. and Wilkinson, C. (2009) 'Opening the Black Box: Scientists' Views on the Role of the News Media in the Nanotechnology Debate', *Public Understanding of Science*, 18 (5), 512–530.

Petts, J., Horlick-Jones, T. and Murdock, G. (2001) *Social Amplification of Risk: The Media and the Public, Contract Research Report*. London: Health and Safety Executive.

PEW (2013) 'Climate Change: Key Data Points from Pew Research'. 5 November. Available at: http://www.pewresearch.org/key-data-points/climate-change-key-data-points-from-pew-research/ accessed 22 November 2013.

PEW (2012) 'Trends in News Consumption: 1991–2012. In Changing News Landscape, even Television is Vulnerable', 27 September. Available at: http://www.people-press.org/files/legacy-pdf/2012%20News%20Consumption%20Report.pdf accessed 17 September 2013.

PEW Project for Excellence in Journalism (2010) '100 Days of Gushing Oil: Eight Things to Know about how the Media covered the Gulf Disaster', 25 August. Available at: http://www.journalism.org/analysis_report/100_days_gushing_oil accessed 10 May 2012.

PEW Project for Excellence in Journalism (2010) *The State of the News Media: An Annual Report on American Journalism*. Available at: http://stateofthemedia.org/2010/ accessed 22 September 2013.

Phillips, A. (2013) 'Journalists as Unwilling Sources'. In Franklin, B. and Carlson, M. (eds.) *Journalists, Sources and Credibility: New Perspectives*. London: Routledge, 49–60.

Phillips, L. (2012) 'Anarchists Attack Science: Armed Extremists are Targeting Nuclear and Nanotechnology Workers', *Nature*, 28 May. Available at: http://www.nature.com/news/anarchists-attack-science-1.10729 accessed 6 November 2013.

Phillips, M. (2010) 'BP's Photo Blockade of the Gulf Oil Spill', *Newsweek*, 25 May. Available at: http://www.thedailybeast.com/newsweek/2010/05/26/the-missing-oil-spill-photos.html accessed 3 May 2012.

Philo, G. (ed.) (1999) *Message Received*. Harlow: Longman.

Philo, G. and Happer, C. (2013) *Communicating Climate Change and Energy Security: New Methods in Understanding Audiences*. London: Routledge.

Pickerill, J. (2006) 'Radical Politics on the Net', *Parliamentary Affairs*, 59 (2), 266–282.

Pickerill, J. (2004) 'Rethinking Political Participation: Experiments in Internet Activism in Australia and Britain'. In Gibson, R., Roemmele, A. and Ward, S. (eds.) *Electronic Democracy: Mobilisation, Organisation and Participation via New ICTs*. London: Routledge.

Pickerill, J. (2003) *Cyberprotest: Environmental Activism Online*. Manchester: Manchester University Press.

Pickerill, J. (2001) 'Weaving a Green Web: Environmental Protest and Computer Mediated Communication in Britain'. In Webster, F. (ed.) *Culture and Politics in the Information Age*. London: Routledge, 142–166.

Pidgeon, N. F. (2012) 'Climate Change Risk Perception and Communication: Addressing a Critical Moment?', *Risk Analysis*, 32 (6), 951–956.

Pidgeon, N. F., Harthorn, B., Bryant, K. and Rogers-Hayden, T. (2009) 'Deliberating the Risks of Nanotechnology for Energy and Health Applications in the US and UK', *Nature Nanotechnology*, 4, 95–98.

Pidgeon, N., Harthorn, B. and Satterfield, T. (2011) 'Nanotechnology Risk Perceptions and Communication: Emerging Technologies, Emerging Challenges', *Risk Analysis*, 31 (11), 1694–1700.

Plunkett, J. (2007) 'BBC Drops Planet Relief After Impartiality Fears', *The Guardian*, 6 September. Available at: http://www.theguardian.com/media/2007/sep/06/television.bbc

Priest, S. H. (ed.) (2012) *Nanotechnology and the Public: Risk Perception and Communication*. New York: Taylor and Francis.

Priest, S. H. (2008) 'North American Audiences for News of Emerging Technologies: Canadian and US Responses to Bio- and Nanotechnologies', *Journal of Risk Research*, 11 (7), 877–889.

Priest, S. H. (1994) 'Structuring Public Debate on Biotechnology: Media Frames and Public Response', *Science Communication*, 16 (2), 166–179.

Priest, S. H., Lane, T., Greenhalgh, T., Hand, L. J. and Kramer, V. (2011) 'Envisioning Emerging Nanotechnologies: A Three-Year Panel Study of South Carolina Citizens', *Risk Analysis*, 31 (11), 1718–1733.

Project for Improved Environmental Coverage (2013) 'Environmental Coverage in the Mainstream News: We Need More'. Available at: http://greeningthemedia.org/wp-content/uploads/Environmental-Coverage-in-the-Mainstream-News.pdf accessed 23 November 2013.

Ragas, M. W., Tran, H. L. and Martin, J. A. (2014) 'A Study of Online Agenda-Setting Effects During the BP Oil Disaster', *Journalism Studies*, 15 (1), 48–63.

Ratter, B. M. W., Philipp, K. H. I. and von Storch, H. (2012) 'Between Hype and Decline – Recent Trends in Public Perception of Climate Change', *Environmental Science & Policy*, 18, 3–8.

Reese, S. D. (2001) 'Prologue - Framing Public Life: A Bridging Model for Media Research'. In Reese, S. D., Gandy, O. and Grant, A. (eds.) *Framing Public Life: Perspectives on Media and our Understanding of the Social World*. Mahwah, NJ: Erlbaum, 7–31.

Reich, Z. (2013) 'Source Credibility as a Journalistic Work Tool'. In Franklin, B. and Carlson, M. (eds.) *Journalists, Sources and Credibility: New Perspectives*. London: Routledge, 19–36.

Reporters without Borders (2009) *Call to Action to Protect Environmental Journalists*, 11 December. Available at: http://en.rsf.org/call-to-action-to-protect-11-12-2009,35315.html accessed 1 December 2012.

Restore the Gulf (2010) Available at: http://www.restorethegulf.gov/release/2010/09/27/submit-suggestion accessed 13 November 2013.

Revkin, A. (2011) 'Climate News Snooze?', *New York Times*, 5 January.

Revkin, A. (2005) 'Bush Aide Softened Greenhouse Gas Links to Global Warming,' *New York Times*, 8 June. Available at: http://www.nytimes.com/2005/06/08/politics/08climate.html?pagewanted=all&_r=0

Robbins, M. (2012) 'Epic Shell PR Fail? No, the Real Villains Here are Greenpeace. Since When Were Greenpeace the Bad Guys?', *The New Statesman*, 18 July. Available at: http://www.newstatesman.com/blogs/sci-tech/2012/07/epic-shell-pr-fail-no-real-villains-here-are-greenpeace accessed 18 November 2012.

Rogers, E. M. and Dearing, W. J. (1988) 'Agenda-Setting Research: Where Has It Been, Where Is It Going? '. In Anderson, J. (ed.) *Communication Yearbook*, 11, Newbury Park, CA: Sage, 555–594.

Rogers-Hayden, T. and Pidgeon, N. F. (2007) 'Moving Engagement Upstream? Nanotechnologies and the Royal Society and Royal Academy of Engineering's inquiry', *Public Understanding of Science*, 16, 345–364.

Rogers-Hayden, T. and Pidgeon, N. F. (2008) 'Developments in Nanotechnology Public Engagement in the UK: "Upstream" towards Sustainability?', *Journal of Cleaner Production*, 16 (8–9), 1010–1013.

Rolfe-Redding, J. C., Maibach, E. W., Feldman, L. and Leiserowitz, A. (2011) *Republicans and Climate Change: An Audience Analysis of Predictors for Belief and Policy Preferences*, Working paper. Available at SSRN: http://ssrn.com/abstract= 2026002 or http://dx.doi.org/10.2139/ssrn.2026002 accessed 4 August 2013.

Routledge, P. (2010) 'Nineteen Days in April: Urban Protest and Democracy in Nepal', *Urban Studies*, 46 (6), 1279–1299.

Royal Academy of Engineering (2009b) *Synthetic Biology: Scope, Applications and Implications*. London: The Royal Academy of Engineering.

Royal Academy of Engineering (2009a) *Synthetic Biology: Public Dialogue on Synthetic Biology*. London: The Royal Academy of Engineering.

Runge, K. K., Yeo, S. K., Cacciatore, M., Scheufele, D. A., Brossard, D., Xenos, M., Anderson, A., Choi, D. H., Kim, J., Li, N., Liang, X., Stubbings, M. and Su, L. Y. F. (2013) 'Tweeting Nano: How Public Discourses About Nanotechnology Develop in Social Media Environments', *Journal of Nanoparticle Research*, 15, 1381.

Russell, A. (2011) *Networked: A Contemporary History of News in Transition*. Cambridge: Polity.

Ryan, C. (1991) *Prime Time Activism: Media Strategies for Grassroots Organizing*. Boston, MA: South End Press.

Sachsman, D. (2000) 'The Role of the Mass Media in Shaping Public Perceptions and Awareness of Environmental Issues,' Climate Change Communication Conference, Ontario, Canada, 22–24 June. In Scott, D. et al. (eds.) *Climate Change Communication: Proceedings of an International Conference*. Waterloo: University of Waterloo. Available at: http://dsp-psd.pwgsc.gc.ca/Collection/ En56-157-2000E.pdf accessed 10 February 2008.

Safford, T. G., Ulrich, J. D. and Hamilton, L. C. (2012) 'Public Perceptions of the Response to the Deepwater Horizon Oil Spill: Personal Experiences, Information Sources, and Social Context', *Journal of Environmental Management*, 113, 31–39.

Salaverría, R. (2002) 'Facts and Trends of Offline and Online Newspaper Publications in Spain: The Impact of the Internet on the Mass Media in Europe', *COST A.20*, Nicosia (Cyprus), March.

Sampei, Y. and Aoyagi-Usui, M. (2009) 'Mass-Media Coverage, its Influence on Public Awareness of Climate-Change Issues, and Implications for Japan's National Campaign to Reduce Greenhouse Gas Emissions', *Global Environmental Change*, 19 (2), 203–212.

Sample, I. (2007) 'Scientists offered Cash to Dispute Climate Study', *The Guardian*, 2 February. Available at: http://www.guardian.co.uk/environment/2007/feb/ 02/frontpagenews.climatechange accessed 10 February 2008.

Satterfield, T., Conti, J., Herr Harthorn, B., Pidgeon, N. and Pitts, A. (2013) 'Understanding Shifting Perceptions of Nanotechnologies and Their Implications for Policy Dialogues About Emerging Technologies', *Science and Public Policy*, 40 (2), 247–260.

Satterfield, T., Kandlikar, M., Beaudrie, C. E. H., Conti, J. and Herr Harthorn, B. (2009) 'Anticipating the Perceived Risk of Nanotechnologies', *Nature Nanotechnology*, 4 (11), 752–758.

Scarce, R. (1990) *Eco-warriors: Understanding the Radical Environmental Movement*. Chicago: Noble Press.

Schäfer, M. S. (2012) 'Online Communication on Climate Change and Climate Politics: A Literature Review', *Wiley Advanced Interdisciplinary Reviews Climate Change*, 3 (6), 527–543.

Schäfer, M. S., Ivanova, A. and Schmidt, A. (2011) 'Global Climate Change, Global Public Sphere? Media Attention for Climate Change in 23 Countries', *Studies in Communication/Media*, 131–148.

Scheufele, D. A. (2006) 'Messages and Heuristics: How Audiences Form Attitudes About Emerging Technologies'. In Turney, J. (ed.) *Engaging Science: Thoughts, Deeds, Analysis and Action*. London: The Wellcome Trust, 20–25.

Scheufele, D. A. (1999) 'Framing as a Theory of Media Effects', *Journal of Communication*, 29, 103–123.

Scheufele, D. A., Corley, E. A., Shin, T. J., Dalrymple, K. E. and Ho, S. S. (2009) 'Religious Beliefs and Public Attitudes Toward Nanotechnology in Europe and the United States', *Nature Nanotechnology*, 4, 91–94. Available at: http://www.nature.com/nnano/journal/v4/n2/abs/nnano.2008.361.html

Schlesinger, P. (1990) 'Rethinking the Sociology of Journalism: Source Strategies and the Limits of Media-Centrism'. In Ferguson, M. (ed.) *Public Communication: The New Imperatives*. London: Sage, 61–83.

Schmidt, M., Torgersen, H., Ganguli-Mitra, A., Kelle, A., Deplazes, A. and Biller-Andorno, N. (2008) 'SYNBIOSAFE E-Conference: Online Community Discussion on the Societal Aspects of Synthetic Biology', *Systems and Synthetic Biology*, 2, 7–17.

Schultz, F., Kleinnijenhuis, J., Oegema, D., Utz, S. and van Atteveldt, W. (2012) 'Strategic Framing in the BP Crisis: A Semantic Network Analysis of Associative Frames', *Public Relations Review*, 38 (1), 97–107.

Schummer, J. (2005) 'Reading Nano: The Public Interests in Nanotechnology as Reflected in Purchase Patterns of Books', *Public Understanding of Science*, 14, 163–183.

Schütz, H. and Wiedemann, P. M. (2008) 'Framing Effects on Risk Perception of Nanotechnology', *Public Understanding of Science*, 17, 369–379.

Sciencewise (2013) *Public Views on Synthetic Biology*. Available at: http://www.sciencewise-erc.org.uk/cms/what-the-public-say-3/?stage=Live accessed 4 October 2013.

Segerberg, A. and Bennett, W. L. (2011) 'Social Media and the Organization of Collective Action: Using Twitter to Explore the Ecologies of Two Climate Change Protests', *The Communication Review*, 14, 197–215.

SEJ (2010) SEJ Letter of 4 June, 2010, to Coast Guard on Media Access, Society of Environmental Journalists. Available at: http://www.sej.org/initiatives/freedom-information/sej-letter-coast-guard-media-access-spill-response-operations accessed 13 November 2013.

Selin, C. (2008) 'The Sociology of the Future: Tracing Stories of Technology and Time', *Sociology Compass*, 2 (6), 1878–1895.

Selin, C. (2007) 'Expectations and the Emergence of Nanotechnology', *Science, Technology and Human Values*, 32 (2), 196–220.

Severin, W. J. and Tankard, J. R. (2001) *Communication Theories: Origins, Methods, and Uses in the Mass Media*. New York: Addison Wesley Longman.

Shanahan, M. (2011) *Why the Media Matters in a Warming World: A Guide for Poli-cymakers in the Global South.* Available at: http://pubs.iied.org/pdfs/G03119.pdf accessed 20 November 2013.

Shanahan, M. (2009) 'Time to Adapt? Media Coverage of Climate Change in Non-Industrialised Countries'. In Boyce, T. and Lewis, J. (eds.) *Climate Change and the Media.* Oxford: Peter Lang, 145–157.

Shanahan, M. (2007) *Talking About a Revolution: Climate Change and the Media, An International Institute for Environment and Development Briefing.* IIED, December.

Shehata, A. and Strömbäck, J. (2013) 'Not (Yet) a New Era of Minimal Effects: A Study of Agenda Setting at the Aggregate and Individual Levels', *International Journal of Press Politics*, 18, 234–255.

Shirky, C. (2011) 'The Political Power of Social Media: Technology, the Public Sphere, and Political Change', *Foreign Affairs,* January/February Issue. Available at: http://www.foreignaffairs.com/articles/67038/clay-shirky/the-political-power-of-social-media accessed 31 October 2013.

Shoemaker, P. J., Vos, T. P. and Reese, S. D. (2009) 'Journalists as Gatekeepers'. In Wahl-Jorgensen, K. and Hanitzsch, T. (eds.) *Handbook of Journalism Studies.* New York: Routledge, 73–87.

Shum, R. V. (2012) 'Effects of Economic Recession and Local Weather on Climate Change Attitudes', *Climate Policy*, 12 (1), 38–49.

Sieber, R. E., Spitzberg, D., Moffatt, H., Brewer, K., Fleki, B. and Arbit, N. (2006) 'Influencing Climate Change Policy: Environmental Non-Governmental Organizations (ENGOs) Using Virtual and Physical Activism', McGill School of Environment, McGill University, Montreal, Canada. Available at: http://www.mcgill.ca/files/mse/ClimateChangePolicyENVR401.pdf

Siegrist, M., Wiek, A., Helland, A. and Kastenholz, H. (2007) 'Risks and Nanotechnology: The Public is more Concerned than Experts and Industry', *Nature Nanotechnology*, 2, 67.

Sigal, L. (1986) 'Sources Make the News'. In Manoff, R. K. and Schudson, M. (eds.) *Reading the News.* New York: Pantheon Books, 9–37.

Skytruth (2010) 'BP Spill: Using Science to Hold BP and Federal Regulators Accountable.' Available at: http://skytruth.org/projects/bp_oilspill/ accessed 13 November 2013.

Smerecnik, K. R. and Renegar, V. R. (2010) 'Capitalistic Agency: The Rhetoric of BP's Helios Power Campaign', *Environmental Communication*, 4 (2), 152–171.

Smith, A. (2010) 'BP's Television Ad Blitz', *CNN Money*, 4 June. Available at: http://money.cnn.com/2010/06/03/news/companies/bp_hayward_ad/?postversion=2010060321 accessed on 15 May 2012.

Speck, D. L. (2010) 'A Hot Topic? Climate Change Mitigation Policies, Politics, and the Media in Australia', *Human Ecology Review*, 17 (2), 125–134.

Stephanian, A. (2010) 'BP's Greenwashing can't Clean Up their Spill', *Huffington Post*, 20 May. Available at: http://www.huffingtonpost.com/andy-stepanian/bps-greenwashing-cant-cle_b_581253.html accessed 11 May 2012.

Straessle, B. (2013) 'Polls Show Voters Strongly Support Offshore Drilling in the US'. Available at: http://www.api.org/news-and-media/news/newsitems/2013/oct-2013/~/media/Files/News/2013/13-October/What-America-Is-Thinking-Access-US.pdf accessed 14 November.

Street Giant (2010) 'Leroy Stick – the Man behind @BPGlobalPR', 2 June. Available at: http://streetgiant.com/2010/06/02/leroy-stick-the-man-behind-bpglobalpr/ accessed 11 May 2012.

Synth-Ethics (2010) 'Synth-Ethics: Identification of Ethical Issues and Analysis of Public Discourse', Report on the first work package. Available at: http://www.synthethics.eu/documents/REPORT%20WP1%20synthethics%20-%20ethics+public%20discourse.pdf accessed 8 September 2013.

Tagbo, E. (2010) *Media Coverage of Climate Change in Africa: A study of Nigeria and South Africa*. Oxford: Reuters Foundation.

Tait, J. (2009) *Upstream Engagement and the Governance of Science: The Shadow of the Genetically Modified Crops Experience in Europe*. EMBO Report, 10, S18-S22 Available at: http://www.nature.com/embor/journal/v10/n1s/pdf/embor2009138.pdf

Takahashi, B. (2011) 'Framing and Sources: A Study of Mass Media Coverage of Climate Change in Peru during the V ALCUE', *Public Understanding of Science*, 20 (4), 543–557.

Takahashi, B. and Meisner, M. (2013) 'Climate Change in Peruvian Newspapers: The Role of Foreign Voices in a Context of Vulnerability', *Public Understanding of Science*, 22 (4), 427–442.

Technorati (2008) 'State of the Blogosphere, 2008'. Available at: http://technorati.com/blogging/state-of-the-blogosphere/ accessed 20 September 2009.

Te Kulve, H. (2006) 'Evolving Repertoires: Nanotechnology in Daily Newspapers in the Netherlands', *Science as Culture*, 15 (4), 367–382.

Telofski, R. (2010) 'Why Social Media will Never Let BP Sleep', 9 July. Available at: http://www.triplepundit.com/2010/07/rebrand-bp-logo-social-media-greenpeace/ accessed 3 May 2012.

Ten Eyck, T. A. and Williment, M. (2003) 'The National Media and things Genetic: Coverage in the New York Times (1971–2001) and the Washington Post (1977–2001)', *Science Communication*, 25 (2), 129–152.

This Week (2010) 'How BP is Controlling Google Results', 7 June. Available at: http://theweek.com/article/index/203754/how-bp-is-controlling-google-results accessed 15 May 2012.

Thorsen, E. (2009) 'Blogging the Climate Change Crisis from Antarctica'. In Allan, S. (ed.) *Citizen Journalism: Global Perspectives*. New York: Peter Lang, 107–120.

Thrall, A. T., Lollio-Fakhreddine, J., Berent, J., Donnelly, L., Herrin, W., Paquette, Z., Wenglinski, R. and Wyatt, A. (2008) 'Star Power: Celebrity Advocacy and the Evolution of the Public Sphere', *International Journal of Press/Politics*, 13, 362–385.

Tollefson, J. (2010) 'Climate Science: An Erosion of Trust?', *Nature*, 466, 24–26. Available at: http://www.nature.com/news/2010/100630/full/466024a.html

Torgersen, H. and Hampel, J. (2012) 'Calling Controversy: Assessing Synthetic Biology's Conflict Potential', *Public Understanding of Science*, 21 (2), 134–148.

Torgersen, H. and Hampel, J. (2001) 'The Gate-Resonance Model. The Interface of Policy, Media and the Public in Technology Conflicts' (December 2001). Institute of Technology Assessment (ITA) Working Paper No. ITA-01-03. Available at SSRN: http://ssrn.com/abstract=475523 accessed 23 November 2013.

Torgersen, H. and Schmidt, M. (2013) 'Frames and Comparators: How Might a Debate on Synthetic Biology Evolve?', *Futures*, 48, 44–54.

Tracy, T. (2010) 'BP Tripled its Ad Budget after Oil Spill', *Wall Street Journal*, 1 September. Available at: http://online.wsj.com/news/articles/SB10001424052748703882304575465683723697708 accessed 13 November 2013.

Tran, H. (2013) 'Online Agenda Setting: A New Frontier for Theory Development'. In Johnson, T. J. (ed.) *Agenda Setting in a 2.0 World: New Agendas in Communication.* London: Taylor and Francis, 205–299.

Trumbo, C. (1996) 'Constructing Climate Change: Claims and Frames in US News Coverage of an Environmental Issue', *Public Understanding of Science*, 5 (3), 269–283.

Tufekci, Z. (2013) ' "Not This One": Social Movements, the Attention Economy, and Microcelebrity Networked Activism', *American Behavioral Scientist*, 57 (7), 848–870.

Tulloch, J. and Lupton, D. (2001) 'Risk, the Mass Media and Personal Biography: Revisiting Beck's Knowledge, Media and Information Society', *European Journal of Cultural Studies*, 4 (1), 5–27.

Turow, J. (2012) *The Daily You: How the New Advertising Industry Is Defining your Identity and your Worth.* New Haven, CT: Yale University Press.

Ungar, S. (1992) 'The Rise and (Relative) Decline of Climate Change as a Social Problem', *The Sociological Quarterly*, 33 (4), 483–501.

Usher, N. (2010) 'Goodbye to the News: How Out of Work Journalists Assess Enduring News Values and the New Media Landscape', *New Media and Society*, 12, 911–928.

Vandermoere, F., Blanchemanche, S., Bieberstein, A., Marette, S. and Roosen, J. (2011) 'The Public Understanding of Nanotechnology in the Food Domain: The Hidden Role of Views on Science, Technology, and Nature', *Public Understanding of Science*, 20 (2), 195–206.

Van Dijck, J. (2013) *The Culture of Connectivity: A Critical History of Social Media.* Oxford: Oxford University Press.

van Dijk, J. A. (1991) *De Netwerkmaatschappij, Sociale Aspecten Van Nieuwe Media.* Houten NL, Zaventem BEL: Bohn Stafleu van Loghum.

Veltri, G. A. (2012) 'Microblogging and Nanotweets: Nanotechnology on Twitter', *Public Understanding of Science*, 22 (7), 832–849.

Vidal, J. (2014) 'MPs Criticise BBC for 'False Balance' in Climate Change Coverage', *The Guardian*, 2 April.

Vidal, J. (2009) 'Copenhagen Media Coverage: A Perfect Storm', *The Guardian*, 7 December.

Vilas Paz, A. (2004) *Prestige: Never Again.* Available at: http://www.archipelago.gr/defaulten.asp accessed 18 May 2005.

Voice of America (2014) Number of Internet Users Worldwide Approaching 3 Billion, 6 May. http://www.voanews.com/content/number-of-internet-users-worldwide-approaching-3-billion/1908968.html

Ward, B. and Hicks, N. (2013) *Submission to inquiry on 'Climate: Public Understanding and Policy Implications' by the House of Commons Select Committee on Science and Technology.* Policy Paper, Grantham Research Institute on Climate Change and the Environment, London. Available at: http://www.lse.ac.uk/GranthamInstitute/publications/Policy/docs/

PP-inquiry-climate-public-understanding-policy-house-of-commons.pdf accessed 22 September 2013.

Warren, J. (2010) *Grassroots Mapping: Tools for Participatory and Activist Cartography*. MSc Thesis, Massachusetts Institute of Technology. Available at: http://unterbahn.com/thesis-web/thesis.pdf accessed 11 November 2013.

Washbourne, N. (2001) 'Information Technology and New Forms of Organising? Translocalism and Networks in Friends of the Earth'. In Webster, F. (ed.) *Culture and Politics in the Information Age*. London: Routledge, 129–141.

Weart, S. R. (2008) *The Discovery of Global Warming*. Cambridge, MA: Harvard University Press.

Weaver, D. A., Lively, E. and Bimber, B. (2009) 'Searching for a Frame: News Media Tell the Story of Technological Progress, Risk, and Regulation', *Science Communication*, 31,139–166.

Weingart, P., Engels, A. and Pansegrau, P. (2000) 'Risks of Communication: Discourses on Climate Change in Science, Politics and the Mass Media', *Public Understanding of Science*, 9, 261–283.

Wheelwright, J. (1994) *Degrees of Disaster: Prince William Sound, How Nature Reels and Rebounds*. New York: Simon and Schuster.

Wickman, C. (2014) 'Rhetorical Framing in Corporate Press Releases: The Case of British Petroleum and the Gulf Oil Spill', *Environmental Communication*, 8 (1), 3–20.

Wilkins, L. (1993) 'Between the Facts and Values: Print Media Coverage of the Greenhouse Effect 1987–1990', *Public Understanding of Science*, 2, 71–84.

Wilkins, L. and Patterson, P. (1991) 'Science as Symbol: The Media Chills the Greenhouse Effect.' In Wilkins, L. and Patterson, P. (eds.) *Risky Business: Communicating Issues of Science, Risk and Public Policy*. Wesport, CT: Greenwood, 159–176.

Wilkinson, C., Allan, S., Anderson, A. and Petersen, A. (2007) 'From Uncertainty to Risk?: Scientific and News Media Portrayals of Nanoparticle Safety', *Health, Risk and Society*, 9 (2), 145–157.

Willard, T. (2009) *Social Networking and Governance for Sustainable Development*. International Institute for Sustainable Development. Available at: http://www.iisd.org/pdf/2009/social_net_gov.pdf accessed 30 October 2013.

Wilson, A. (1992) *The Culture of Nature: North American Landscape from Disney to the Exxon Valdez*. Cambridge, MA: Blackwell.

Wolfsfeld, G. (1997) *Media and Political Conflict: News from the Middle East*. Cambridge: Cambridge University Press.

World Bank (2013) 'What Climate Change means for Africa, Asia and the Coastal Poor', 19 June. Available at: http://www.worldbank.org/en/news/feature/2013/06/19/what-climate-change-means-africa-asia-coastal-poor accessed 16 November 2013.

World Wildlife Fund (2002) *The Prestige Catastrophe*. November 2002 - Oil spill off Spain's NW coast, Available at: http://www.panda.org/news_facts/crisis/spain_oil_spill/

Wu, Y. (2009) 'The Good, the Bad, and the Ugly: Framing of China in News Media Coverage of Global Climate Change'. In Boyce, T. and Lewis, J. (eds.) *Climate Change and the Media*, 158–173. Oxford: Peter Lang.

Xie, L. (2011a) *Environmental Activism in China*. London: Routledge.

Xie, L. (2011b) ' China's Environmental Activism in the Age of Globalization', *Asian Politics and Policy*, 3 (2), 205–222.

YouGov (2013) 'Carbon Trust Results', 12 September 2013 Environmental Sustainability. Available at: http://cdn.yougov.com/cumulus_uploads/document/b3dc03mvhx/YG-Archive-Carbon-Trust-results-120913-environmental-sustainability.pdf accessed 14 November 2013.

YouGov (2011) 'One Year after BP Oil Spill Americans are Ready to Drill', 22 April. Available at: http://today.yougov.com/news/2011/04/22/one-year-after-bp-oil-spill-americans-are-ready-dr/ accessed 14 November 2013.

Young, N. and Dugas, E. (2011) 'Representations of Climate Change in Canadian National Print Media: The Banalization of Global Warming', *Canadian Review of Sociology*, 48 (1), 1–22.

Zehr, S. C. (2000) 'Public Representations of Scientific Uncertainty of Global Warming', *Public Understanding of Science*, 9 (2), 85–103.

Zhang, J. Y., Marris, C. and Rose, N. (2011) 'The Transnational Governance of Synthetic Biology: Scientific Uncertainty, Cross-Borderness and the "Art" of Governance'. BIOS Working Paper No: 4. London: LSE. Available at: http://www.kcl.ac.uk/sspp/departments/sshm/research/csynbi/TransnationalGovernanceofSyntheticBiology.pdf accessed 18 October 2012.

Zimmer, R., Hertel, R. and Böl, G. -F. (eds.) (2010) *Risk Perception of Nanotechnology – Analysis of Media Coverage*. Berlin: Federal Institute for Risk Assessment. Available at: http://www.bfr.bund.de/cm/350/risk_perception_of_nanotechnology_analysis_of_media_coverage.pdf accessed 16 September 2013.

Zimmer, R., Hertel, R. and Böl, G. -F. (2008) *Public Perceptions About Nanotechnology: Representative Survey and Basic Morphological-Psychological Study*. Berlin: Federal Institute for Risk Assessment. Available at: http://www.bfr.bund.de/cm/350/public_perceptions_about_nanotechnology.pdf accessed 22 September 2013.

Index

Printed and bound in the United States of America